Library of
Davidson College

ON PICTURE VARNISHES AND THEIR SOLVENTS

A publication based on the principal papers presented at a *Seminar on Resinous Surface Coatings* sponsored by the Intermuseum Conservation Association, Oberlin, Ohio.

ON PICTURE VARNISHES AND THEIR SOLVENTS

REVISED AND ENLARGED EDITION

Robert L. Feller
Nathan Stolow
Elizabeth H. Jones

NATIONAL GALLERY OF ART, WASHINGTON

Original publication Copyright © 1959 Intermuseum
Conservation Association, Oberlin, Ohio.

Revised edition Copyright © 1971 The Press
of Case Western Reserve University, Cleveland, Ohio 44106.
This revised and enlarged edition was made possible
by a grant from the Samuel H. Kress Foundation.

Copyright © 1985 Board of Trustees, National Gallery of Art,
Washington. All rights reserved. No part of this publication
may be reproduced without written permission of the
National Gallery of Art, Washington, D.C. 20565.

COVER: Jean-Baptiste-Siméon Chardin, *The House of Cards*,
c.1735, National Gallery of Art, Washington, Andrew W.
Mellon Collection 1937.1.90. Detail showing partial removal
of varnish.

Library of Congress Cataloging in Publication Data

Feller, Robert L.
  On picture varnishes and their solvents.

  Reprint. Originally published: Cleveland : Press
of Case Western Reserve University, 1971
  "Based on the principal papers presented at a
Seminar on Resinous Surface Coatings sponsored by the
Intermuseum Conservation Association, Oberlin, Ohio"—
  Includes bibliographies and index.
  1. Varnish and varnishing. 2. Artists' materials.
I. Stolow, Nathan.   II. Jones, Elizabeth, H.   III. Title.
ND1530.F4   1985   751.6   85-7172
ISBN 0-89468-084-6

CONTENTS

Foreword to 1985 Edition	ix
Foreword to 1971 Edition	x
Foreword to the First Edition	xii
Preface to 1971 Edition	xiv
Preface to the First Edition	xvi
Authors' Acknowledgments	xxi

PART I Solvents *Robert L. Feller*	1
1. DESCRIPTION OF SOLVENT-TYPE VARNISH	3
The First Period in the Life of a Film	4
The Second Period in the Life of a Film	5
The Third Period in the Life of a Film	5
References and Notes	6
2. FUNDAMENTAL ORGANIC COMPOUNDS	7
Hydrocarbons	8
Petroleum Fractions	11
Turpentine	13
Unsaturated Hydrocarbons	14
Derivatives of Paraffin Hydrocarbons: Alcohols	14
Acids, Esters, Ketones, and Ethers	17
Organic Compounds Containing Halogens and Nitrogen	21
Summary	21
References and Notes	22

3. VOLATILE SOLVENTS	24
Solution	24
Chemical type	25
Volatility	25
Solvent Action	26
Safety	38
Summary	40
References and Notes	41

PART II Solvent Action *Nathan Stolow* 45

4. SOLVENT ACTION: SOME FUNDAMENTAL RESEARCHES INTO THE PICTURE-CLEANING PROBLEM	47
Introduction	47
Some Remarks on the Drying of Linseed Oil Films	50
Leaching and Swelling Action of Solvents	54
Initial Swelling/Leaching Process	55
Repeated Swelling on Leached Films	60
Leaching Action and Time of Solvent Contact	62
Amount of Leaching: Solvent and Film Type	63
Amount of Leaching as the Film Ages	66
The Leached Components	68
Infra-red Studies	70
Paper Chromatography	72
Gas Chromatography	73
Swelling Action in Different Solvents	88
Molecular Volume of Solvents	89
Solubility Parameters	90
The Rate of Swelling Action; Diffusion of Solvents	95
General Conclusions	105
Summary	108
References and Notes	111

PART III Resins and the Properties of Varnishes *Robert L. Feller* 117

5. THE NON-VOLATILE COMPONENT	119
Natural Resins	120
Synthetic Thermoplastic Polymers	125
References and Notes	133

6. PROPERTIES OF FRESHLY APPLIED VARNISH	137
Drying of Varnish	137
Control of Appearance	139
References and Notes	144
7. PROPERTIES OF MATURE VARNISH	146
Protection by Varnish	146
1. Protection from Dirt and Abrasion	146
2. Protection During Normal Movement of a Painting	147
3. Protective Value of a Varnish That Requires "Mild" Solvents and Infrequent Removal	147
4. Protection from Radiation	148
5. Considerations Regarding the Permeability of Picture Varnish	150
Specifications for Picture Varnish	152
Deterioration of Varnish	154
References and Notes	165

PART IV The Removal of Varnish *Elizabeth H. Jones* 169

8. INVESTIGATIONS ON THE REMOVAL OF AGED VARNISH COATINGS	171
Tests on the Ease of Removal of Various Varnishes	171
Description of Specimens	171
Testing Procedures	172
Summary of Results	177
The Re-forming of Varnish Coatings	178
The Technique	178
Precedents and History	179
Tests of Re-forming on 1938 Fogg Test Panels	179
Summary of Results	179
Practical Application of the Re-forming Technique	180
Investigations into the Mechanism of Re-forming an Aged Varnish	181
Infra-red Spectroscopy	181
Radioactive Ethanol	185
Summary and Conclusion on the Technique of Re-forming Varnish Coatings	188
References and Notes	189

APPENDICES

A. Early Studies on the Cross-Linking of Polymers
   Robert L. Feller — 195
   References and Notes — 201

B. Solubility and Removability of Aged Polymeric Films
   Robert L. Feller — 202
   Induction Period — 202
   Period of Decreasing Solubility — 203
   Period of Decreasing Swelling — 205
   Period of Non-Swelling by Solvents — 206
   Summary and Conclusions — 206
   References and Notes — 208

C. Studies of the Effect of Light on Protective Coatings Using Aluminum Foil as a Support: Determination of the Ratio of Chain Breaking to Cross-Linking   *Robert L. Feller* — 211
   Summary — 216
   References and Notes — 217

D. Polymer Emulsions   *Robert L. Feller* — 218
   References and Notes — 224

E. Grades of Poly(vinyl Acetate) Resins with Respect to Their Viscosity in Solution   *Robert L. Feller* — 226
   References and Notes — 228

F. Physical, Chemical, and Toxicological Properties of Solvents and Liquids for Conservation of Paintings and Works of Art   *Nathan Stolow* — 230
   Table and Notes — 231

INDEX — 235

BIBLIOGRAPHY — 252

FOREWORD TO 1985 EDITION

Since it was first published in 1959, this book has served as an introductory text on spirit varnishes, describing the inherent properties of thermoplastic resins, their life cycles and how they may be safely removed. An understanding of the basic aspects of the application and utilization of solvent-based coatings is necessary for the artist and conservator if they are to employ with confidence the wide range of natural and synthetic resins available. Although some of these materials are used as adhesives, consolidants, and vehicles for retouching, the text primarily discusses the traditional and modern materials that have been most widely used as protective coatings for easel paintings. The properties of certain polymers such as nitrocellulose, polyvinyl chloride, and polystyrene are not discussed because these materials have been generally considered insufficiently stable for use in conservation.

Although a major revision and extension of the text has not been possible since the 1971 revised edition, the principles underlying the behavior of solvents and thermoplastic resins remain unchanged. We trust that the growing numbers of conservators, artists, curators, and collectors with the need to understand the properties and behavior of resinous coatings will be well served by the reissue of this volume. A brief bibliography of pertinent contributions that have appeared since 1971 is given as an addendum.

The National Gallery of Art initiated this reprinting, but we are most appreciative of the Foundation of the American Institute for Conservation, which is responsible for its distribution.

Ross M. Merrill
Chief of Conservation, National Gallery of Art

FOREWORD TO 1971 EDITION

On Picture Varnishes and Their Solvents, by Feller, Stolow, and Jones, was first published by the Intermuseum Conservation Association in 1959, with the assistance of the Lilly Endowment, Inc. The book proved to be a useful contribution to the expanding knowledge of materials and techniques in the conservation of art, and the first edition was soon exhausted. Yet requests for it continue, and its reissue has been urged by many.

Now, with the generous assistance of the Samuel H. Kress Foundation and of the authors, the Intermuseum Conservation Association is able to comply with the demand and to offer a second printing which, however, has been revised and updated by Dr. Feller, Dr. Stolow, and Miss Jones under the leadership of Mr. Richard D. Buck, Chief Conservator of the Association. We are confident that the revised edition will be even more useful than the first edition and will prove to be a further contribution to our expanding knowledge.

The trustees of ICA, representing the following member museums:

Albright-Knox Art Gallery
Buffalo, New York

Allen Memorial Art Museum
Oberlin, Ohio

The Cleveland Museum of Art
Cleveland, Ohio

The Columbus Gallery of Fine Arts
Columbus, Ohio

Galleries of Cranbrook Academy
　of Art
Bloomfield Hills, Michigan

Davenport Municipal Art Gallery
Davenport, Iowa

The Dayton Art Institute
Dayton, Ohio

Elvehjem Art Center
Madison, Wisconsin

Indianapolis Museum of Art
Indianapolis, Indiana

The University of Michigan
　Museum of Art
Ann Arbor, Michigan

The Minneapolis Institute of Arts
Minneapolis, Minnesota

Munson-Williams-Proctor Institute
Utica, New York

The Memorial Art Gallery
Rochester, New York

The Taft Museum
Cincinnati, Ohio

The Toledo Museum of Art
Toledo, Ohio

FOREWORD TO THE FIRST EDITION

At the time the Intermuseum Conservation Association was founded, certain unusual potentialities inherent in a cooperative activity were foreseen. A number of museums, united by such an association and its laboratory, might expect to gain more than a service in conservation. Accordingly, the following aims were established and appear in its Articles of Incorporation:

> To improve and disseminate knowledge of the theory and practice of conservation in relation to works of art and objects of cultural interest.

> To investigate materials and equipment and to conduct studies and tests in order to develop methods to protect, preserve, maintain the integrity of, and improve the condition of such works of art and objects of cultural interest.

These aims were implemented in 1955 when experimental samples of a new varnish were supplied to the laboratory by the Fellowship on Artists' Materials at the Mellon Institute. The Association agreed to allow certain of its paintings to be treated with this varnish and to observe its appearance under various conditions existing in member museums. By so doing, the Association felt it might contribute a critical study of the material not possible in a laboratory. This field testing is being continued with the technical advice of the Fellowship.

In 1956 a special fund was established by the member museums and by a generous grant from the Lilly Endowment, Inc., to further the corollary activities of the Association in research and education. This fund made it possible for the Intermuseum Conservation Association to offer its Seminar on Resinous Surface Coatings in the spring of 1957. That the seminar was a valued contribution to the understanding of this complex subject is due to the willing cooperation of those who have been instrumental in recent research, and to the Chief Conservator of the Association, Mr. Richard D. Buck, without whose knowledge, insights, and vigorous leadership this seminar would not have borne such fruit. The Intermuseum Conservation Association gratefully acknowledges its debt to the Lilly Endowment, Inc., and to its member museums, and takes pride in publishing an important part of the seminar.

The Trustees of ICA

PREFACE TO 1971 EDITION

Even though the first edition of this book was exhausted some time ago, there has been a continuing demand for it. Concurrent with an increasing concern with problems of conservation, the past decade has seen a wide adoption of polymer materials as coatings and adhesives. This book, summarizing such recent developments in varnish technology, still serves as the solitary liaison between the research chemist and the museum world, in spite of the extensive literature on varnishes and varnishing. To our knowlege no similar work is in print or in preparation.

For this revised edition each author has been asked to add new data from both his own research and references to recent work by other investigators. Thus, the revised edition is enlarged both in text and in references. It is still basically the publication of the 1957 ICA Seminar at Oberlin and does not attempt to be a coordinated compendium on varnish and solvent technology.

To make the book more useful to a growing number of readers, certain practical changes have been made. An index has been added, and printing and binding have been improved. The production of this edition has been in the hands of a professional publisher, and the book is available through normal marketing channels.

ICA's position as coordinator and editor is inherited from the occasion of the seminar. This service is compatible with the declared purposes of the Association. ICA is privileged to have had such admirable cooperation from the three authors, on whose heavy schedules great demands have been made. We are grateful also to The Press of Case Western Reserve University for its helpful counsel.

Finally, ICA wishes to reaffirm its sincere gratitude to the Lilly Endowment, Inc., for financing the printing and distribution of the first edition, and to acknowledge its present indebtedness to the Kress Foundation for making possible the revised edition.

Richard D. Buck
Director, Intermuseum Laboratory

PREFACE TO THE FIRST EDITION

Varnish, being a pure organic material, is understood to be more vulnerable to change and decay than is paint. Paint is given durability by its inorganic component, pigment. Varnish is conceded to be a temporary addition to the structure of a painting. It is a common observation that if the varnish traditionally used on paintings is not periodically replaced, an artist's design will recede, with the passage of time, behind an obscuring film. On the other hand, if aged varnishes are replaced repeatedly, some damage to paint, however slight, must be acknowledged as a consequence of the cleaning process.

The dilemma imposed by the transitory nature of picture varnish has been apparent for centuries. And yet, from what can be read of its history, there has been but one major technical advance in the use of varnish—and that occurred some four hundred years ago. During the late Renaissance, perhaps after a refinement in the art of distillation, spirit varnish began to displace oil varnishes. A clearer varnish, more easily and safely removable, was produced. Except for improvements resulting from better selection of materials and greater control in manufacture, picture varnish has since remained essentially unchanged.

Now we may be about to make another step forward. In the last two or three decades, thanks to the technology underlying our burgeoning plastics industries, we have learned more by far about resins, their nature and behavior, and about solvents and their actions than has ever been known before. There is a gathering momentum in our progress toward another marked improvement in coating materials. Already several new varnishes have been developed that show notable freedom from familiar defects.

But advances rarely come about solely as a result of new technology. Innovations in practice require the cooperation of collectors, curators, and all others responsible for the care of paintings. They must consider the consequence of the employment of new techniques with deliberate caution, avoiding blind conservatism as well as impetuous approval. Only a basic understanding by all concerned of the nature of polymers, of solvents, and of solvent action would make possible a constructive exchange of information.

The Seminar on Resinous Surface Coatings held at Oberlin in April 1957, was arranged to present a summary of present knowledge of the properties and uses of varnishes and lacquers, and of the consequences of their use on paintings. The Seminar was intended for conservators and museum personnel of curatorial responsibility. A large part of the material covered was technical in nature, but all speakers were encouraged to make information understandable to those whose formal scientific training might be elementary.

The papers of three of the lecturers are presented in this publication. Each of these papers is the report of recent original investigation extending over a period of years.

Dr. Robert L. Feller was appointed in 1950 to the Fellowship on Artists' Materials at Mellon Institute, established at that time by the National Gallery of Art, Washington, D.C. His first major investigation was directed toward the development of a picture varnish that would be highly resistant to yellowing, would retain its original flexibility and transparency, and would be removable with mild solvents. This challenging task has required a thorough study of traditional varnishes, modern high polymers, and solvents in order to identify and measure the properties that are sought in a varnish. Thus the report of his research to the present date is almost a text on varnish technology. In preparing his lectures Dr. Feller further agreed to review certain pertinent areas of organic chemistry in such a way that the discussions would be comprehensible to non-chemists. His lectures, therefore, provided the main frame of the seminar.

Dr. Nathan Stolow, Head of the Department of Conservation and Scientific Research at the National Gallery of Canada, Ottawa, reported on investigations carried out at the University of London in preparation for his doctoral dissertation. His explanation of the action of solvents on linseed oil films is a major achievement, providing for the first time a basic understanding of factors that must be given most careful consideration in the manipulation of varnishes.

Miss Elizabeth Jones, Conservator of the Fogg Museum, Harvard University, has made an extremely valuable contribution in her study of solvent retention in varnish films and the removal of aged films, providing explanations for a number of phenomena that previously have been perplexing.

To form a logical sequence in this publication, the papers by Dr. Stolow and Miss Jones have been interpolated as chapters in the longer statement by Dr. Feller.

The association chose to publish only those parts of the Seminar which report new experimental data, yet it gratefully acknowledges the contributions of the other speakers at the Seminar.

Mr. Theodore Rousseau, Jr., Curator of Paintings at the Metropolitan Museum of Art, New York, presented in a public lecture a careful survey of those aesthetic qualities of varnish which contribute to the satisfactory appearance of paintings.

Dr. Peter Hawkins, Assistant Professor of Chemistry, Oberlin College, gave a lucid explanation of fractional distillation and chromatography. These processes, apparently unrelated to the arts, have their application in the refining of petroleum solvents on the one hand, and in the analysis of resinous materials on the other. Demonstrations of distillation and chromatography were arranged by Mr. Peter Michaels, graduate student in Fine Arts, under the instruction of Professor William Renfrow.

Mr. Michaels also presented a brief survey of the history of varnish, with added commentary by Mr. R.J. Gettens of the Freer Gallery, Washington, D.C. Mr. Gerhard Wedekind, Assistant Conservator, Metropolitan Museum of Art, and Miss Elizabeth Jones demonstrated new techniques in cleaning paintings.

A trip to the varnish sheds of the Sherwin-Williams Company in Cleveland to see the production of oil varnishes closed the Seminar. Oil varnish coatings, now displaced by spirit varnishes, were in common use in medieval and Renaissance painting.

Dr. Feller's study of the deterioration of high polymer films has continued. An appendix has been added containing further observations that complement the data available at the time of the Seminar.

We gratefully acknowledge permission of Mr. H. Burrell and his publishers to reproduce as Appendix 2[1] his table of solubility parameters. A third appendix[2] assembled by Dr. Feller relates various grades of poly(vinyl acetate) manufactured in America and Europe. We are glad to print this additional information because of its pertinence to the subject of the seminar.

Among the forty-seven persons who attended the Seminar were twenty-five engaged professionally in conservation and restoration, eight curators, six museum directors, four scientists, one art teacher, and three students. These persons came from fourteen states, Canada, and France.

Two years have elapsed between the Seminar and this publication. ICA did not originally intend to publish the Seminar papers. But a number of those who attended the Seminar pointed out

1. See Table 4-13 and Appendix F.
2. See Appendix E.

that the parts included here would have a special and permanent value if they were published together rather than separately. The final decision to print this material was influenced in a very great degree by the desire to make a single reference work available to all who are concerned with works of art. This publication will be distributed with the compliments of the ICA to the Fellows and the Associates of the International Institute for the Conservation of Museum Objects, whose membership is composed of those especially interested in the problems of conservation.

Beyond a grateful acknowledgment to all of those who contributed to the Seminar, special thanks must be given Dr. Feller, who bore a large burden in the preparation of the manuscripts for printing. The editing, however, has been the responsibility of the staff of the Intermuseum Laboratory, and we are accountable for certain liberties taken to adapt the texts for publication.

R.D.B.

AUTHORS' ACKNOWLEDGMENTS

The National Gallery of Art, Washington, D.C., has maintained a Research Project at Mellon Institute since 1950 through the generosity of the Old Dominion and Avalon foundations. The expressed objective of this research program has been to develop new materials and techniques for the fine arts, both for original work and for conservation, with a special view towards permanence. Dr. John Walker, the first Chief Curator of the National Gallery of Art, was personally responsible for the establishment of the Research Project and its objectives. Later, when Dr. Walker became the Director, he asked Mr. Ernest R. Feidler, Administrator and later Chief Administrative Officer of the National Gallery of Art, to oversee the activities of the research establishment. It is through the outstanding personal interest of both men that the long-term research program at Pittsburgh has been so vigorously supported.

    The craftsman of the past drew upon many generations of experience in the application of his waxes, resins, and adhesives, yet his modern counterpart has had no such advantage with respect to the new materials that chemical industry has developed. Moreover, the conservator's opportunity for experience with acrylic and vinyl acetate resins was delayed by World War II, so much so that when the Research Project was established in 1950 there remained many practical questions to be answered regarding the handling and long-term properties of these materials. Consequently, the Research Project's immediate objective was not to seek solutions to specific problems, but rather to increase the general understanding of resins and their solvents. Because it is impossible for the research and development laboratory to anticipate every variation and problem in usage, the Research Project considered that the introduction of new materials and techniques would be most successful if the craftsman had sufficient understanding of these distinctly new materials so that he himself could control their behavior effectively according to the needs of a particular situation. It is this objective, imparting a broad understanding of these new materials and their behavior, that much of the research and technical information presented at the Oberlin seminar was originally designed to meet. This, then, is a monograph on the chemistry and physics of solvent-type coatings based on thermoplastic resins, particularly as they may be employed as protective coatings on paintings. Directions are not given on how to solve any specific problem; but, hopefully, the Research Project's contributions provide a number of insights through which the conservator may gain increased confidence in the selection of his solvents and resins.

An undertaking of this character and scope is only possible through the thought and energy contributed by many persons. Over the years, a number of chemists have participated in the protective coatings research program. Mr. (now Dr.) Stuart Raynolds, associated with the Research Project as Junior Fellow from 1951 to 1955, contributed the principal data on the physical properties of resins, which were presented formally in a Master of Science thesis at the University of Pittsburgh in 1954, entitled "The Dependence of Physical Properties on the Constitution of Alkyl Polymethacrylic-Ester Polymers." The studies on deterioration were ably conducted for many years by Catherine Westervelt Bailie, followed, for briefer periods of association, by Richard A. Tauson, Joseph J. Matous, Nile Lestrange, and Jean Barr Page. An essential aspect of the successful development of new methods is, of course, the "feedback" to the laboratory of practical experience from the field. Following the development of a number of coating materials in the laboratory, intensive field trials were first made by Mr. Mario Modestini and Mr. Paul Kiehart; since that time much information has been gratefully received from many IIC fellows in the United States regarding their experience in practical application. The Research Project wishes to acknowledge the extensive testing of these materials in the field by Dr. Eric C. Hulmer which began many years before the establishment of the Project. Since 1950 his results have proved especially valuable, and his advice, counsel and exchange of information have been of extremely great value in the practical development of new coating and solvent systems.

Finally, grateful acknowledgment is made to those who have so generously given advice and criticism regarding this and other manuscripts prepared in the course of the work, especially to Professor and Mrs. Sheldon Keck, and Mr. and Mrs. Russell Quandt. Above all, over a period of many years of association, Mr. Richard D. Buck has given valuable and much appreciated criticism. The author, and the reader too, are greatly indebted to him for his labor, thought, and long devotion to the ideas that were embodied in the seminar of 1957 and in the book that was based on the principal lectures, ideas which now find renewed expression in this extended and revised publication.

Robert L. Feller
Carnegie-Mellon University, Pittsburgh, Pennsylvania

In the original edition Chapter 4 was based on Ph.D. researches carried out at the Courtauld Institute of Art, University of London, 1952-56. Acknowledgment was given then to Mr. S. Rees Jones of the Courtauld Institute, under whose general guidance the work was carried out.

In preparing the extensive revisions of Chapter 4, the author acknowledges gratefully the support of Miss Jean S. Boggs,

Director of the National Gallery, in these researches and the excellent assistance in gas chromatography, special examinations, tests and experiments rendered by Mr. George Rogers, Dr. James Hanlan, and Mr. Raymond Boyer, staff scientists of the National Conservation Research Laboratory.

Nathan Stolow
National Gallery of Canada, Ottawa, Canada

The author wishes to express her gratitude to Mr. A. K. Doolittle, Dr. Paul M. Doty, and Dr. Arthur K. Solomon for their comments on theory and their practical advice and help in the investigations reported in Chapter 8.

Dr. David Robinson and Dr. George Hauser have most generously given us their time and graciously responded to repeated calls upon their specialized knowledge. The infra-red and C^{14} ethanol studies would have been impossible without their aid.

The preparation, preservation, and records of the collection of samples used in testing are the work of former members of the Department of Conservation: Messrs. George L. Stout, Rutherford J. Gettens, and Richard D. Buck. The author wishes to thank them for this and for their interest and advice.

In particular, the author wishes to acknowledge her great debt to Mr. Morton C. Bradley, Jr., who first used the re-forming technique systematically as a preparation for cleaning and who contributed so much to this research.

The author is most grateful to her colleagues at the Fogg Museum, Dr. John Coolidge and Mr. Steven Herbert. Without their constant interest and help, the research and this paper could not have been completed.

Elizabeth H. Jones
Fogg Museum of Art, Cambridge, Massachusetts

Part I

SOLVENTS

Robert L. Feller

1
DESCRIPTION OF SOLVENT-TYPE VARNISH

Varnishes are customarily placed in two classifications: solvent-type (or spirit), and oil-resin. A solvent-type varnish consists simply of a volatile solvent and a non-volatile substance; the latter is usually amorphous and thermoplastic, exhibiting relatively little resistance to solution.[1] Such a varnish dries by loss of solvent. Oil-resin varnishes, on the other hand, dry only partially in this manner. Their major mode of drying is distinguished by a chemical change, which may include oxidation and polymerization. Oil-resin coatings are related in character to thermosetting resins which become resistant to solution and thermal softening as the chemical reactions proceed.

The names "spirit varnish," "solvent-type varnish," "solvent-thinned varnish," and "lacquer" are often used interchangeably. The latter name, widely used in industry, is not satisfactory because of the possibility of confusion with oriental lacquer.[2] In the first edition of this book "spirit varnish" was used as a standard term, as is accepted by such authorities as Mattiello and the American Society for Testing Materials.[3] The word "spirit" is not intended to restrict the meaning to ethyl alcohol, but refers to distillation or volatility, as it does in the expressions "mineral spirits" and "spirits of gum turpentine." This older expression is little used today, however, and the term "solvent-type varnish" is more appropriate.

The properties of a solvent-type coating depend ultimately upon the properties of the non-volatile component. In a recently prepared film, however, the properties are sufficiently affected by the solvent that special precautions must be taken before making comparative measurements on the so-called dry film. For reasons of this kind it is appropriate to regard the physical properties of a film of solvent-type varnish as falling into three periods of time with respect to the age of coating: the first, a period after the film has become dry to the touch, during which the concentration of retained solvent is still

4   THE SOLVENTS

rapidly changing; the second, a much longer period, during which there are no readily measurable processes of solvent loss or film degradation; the third, a period during which the variation of properties can be attributed to degradation of the film. Although these periods are arbitrary and are not clearly distinguished during the transition from one into the other, this conceptual approach is useful in emphasizing the variable nature of a film of varnish and organizing the study of properties at various stages in time.

Factors in all three periods are of importance in evaluating the overall performance of a solvent-type varnish. It is the second or middle period, however, in which properties such as tensile strength, flexibility, thermal-distortion temperature, and brittle point are considered to be characteristic of a film of the non-volatile, film-forming substance.

THE FIRST PERIOD IN THE LIFE OF A FILM

The solvent has a transitory and variable effect upon the properties of the film of varnish. This is particularly apparent during the first period described.

A solvent-type varnish becomes dry to the touch long before the solvent has largely evaporated. Typical drying curves are shown in Figure 1-1; the last few per cent of solvent is retained rather tenaciously. We have observed that as

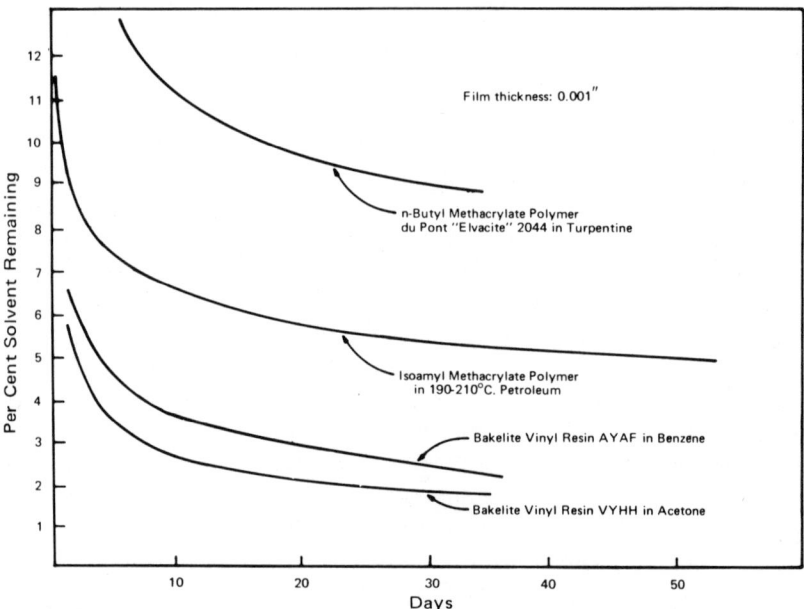

Figure 1-1   Drying Curves of Solvent-type Coatings (at 70°F. and 50% R.H.).

much as 15% of the weight of a film of dammar-turpentine varnish may still be volatilized after it has dried for a month at 70°F. (21.1°C.) and 50% relative humidity. The solvent remaining in the film influences such properties as hardness and brittleness.[4,5]

The choice of solvent cannot be made without some consideration of the non-volatile component; nitrocellulose, for example, will not dissolve in aromatic hydrocarbons. Poly(vinyl acetate) will not dissolve in xylene; poly(n-butyl methacrylate) will. This question of solubility has been an important aspect of the problem of adapting synthetic resins for picture varnish, because many of the synthetic resins first produced by industry required ketones and esters, usually regarded as "strong solvents," for their solution. As our knowledge of synthetic resins has grown, however, the possibility has arisen of selecting and developing resins which are soluble in "mild" or special solvents, whichever may be needed for a particular application.

The formulation of the solvent also influences viscosity, brushability, gloss, blooming or blushing, stresses within the film, and the tendency to discolor. Some of the effects of solvents are significant in the second and third periods in the life of a film. Books on the technology of solvents, such as that of Doolittle, discuss many of these problems,[6,7] Effects that may be altered by suitable formulation should be clearly distinguished from qualities attributable to the resin; it should be recognized, for example, that qualities such as the surface gloss of a mature film are strongly influenced by the formulation of the solvent.[8,9]

THE SECOND PERIOD IN THE LIFE OF A FILM

After the solvent has largely departed, a long period occurs usually in which the properties are reasonably stable. In this second period the properties may be considered to be characteristic of the film-forming, non-volatile component. Attainment of this stage may be hastened in the laboratory by baking the coating to drive off as much of the solvent as possible or by drying it under a vacuum.

In practice, the properties of a varnish will not be those of the thoroughly dry state, but will appear as a range. In the case of dammar and mastic varnishes this range was estimated by the Research Project from two sets of measurements, one made after drying for one month at room conditions and the other made after the films had been baked for two days at 70°C.[5]

THE THIRD PERIOD IN THE LIFE OF A FILM

After the second period which, as has been mentioned, is usually of much longer duration than the first, comes a third period, during which changes in properties occur, due to the deterioration of the film by oxidation and other reactions, often enhanced by heat and light. Discoloration is one such effect; embrittlement and hardening are others.

The physical changes which take place with age, such as discoloration or embrittlement, are frequently traceable to chemical processes occurring within the film. For example, the discoloration of "Saran," poly(vinylidene chloride), has been attributed to the stripping-off of chlorine and hydrogen atoms.[10] Physical processes, such as separation into a two-phase system, syneresis of a gel, or release of stresses, may also be responsible for some of the effects observed in aged films.

For convenience, the volatile and non-volatile components of a solvent-type varnish will be described separately. Bear in mind, however, that each component in the film of varnish may be influenced by the presence of others, in the manner just outlined.

REFERENCES AND NOTES

1. Definitions of the words "resin," "amorphous," and "thermoplastic" are given in Chapter 5.
2. R. J. Gettens and G. L. Stout, *Painting Materials* (New York: Van Nostrand, 1942), p. 31.
3. F. Damitz and J. A. Murphy, "Varnishes," in J. J. Mattiello (ed.), *Protective and Decorative Coatings* (New York: Wiley, 1943), 3:195.
4. R. L. Feller, "Hardness and Flexibility of Natural and Synthetic-resin Varnishes," *Museum News* 29, No. 20 (1952): 7-8; see also R. Katz and B. F. Munk, "The Influence of the Solvent and of the Substrate on the Water Vapor Permeability of Films," *J. Oil & Colour Chemists' Assoc.* 52 (1969): 418-32.
5. R. L. Feller, "Dammar and Mastic Varnishes—Hardness, Brittleness and Change in Weight Upon Drying," *Studies in Conservation* 3 (1958): 162-74.
6. A. K. Doolittle, *The Technology of Solvents and Plasticizers* (New York: Wiley, 1954).
7. W. W. Reynolds, *Physical Chemistry of Petroleum Solvents* (New York: Reinhold, 1963).
8. H. Schubart, "Failure of Synthetic Materials in Picture Conservation," *Museums J.* 53 (1953): 239-40.
9. R. L. Feller, "Factors Affecting the Appearance of Picture Varnish," *Science* 125 (1957): 1143-44.
10. L. A. Matheson and R. F. Boyer, "Light Stability of Polystyrene and Polyvinylidene Chloride," *Ind. Eng. Chem.* 44 (1952): 867-74.

2
FUNDAMENTAL ORGANIC COMPOUNDS

The materials encountered in solvent-type varnishes are generally derived from the class of substances known as organic compounds, as opposed to inorganic compounds. The chemistry of organic substances was once considered to be in the realm of those materials which could arise only through the operation of a vital force inherent in a living cell. The doctrine of an essential vital force (the doctrine that there was something fundamental and distinctive about the substances found in nature), remained unchallenged until 1828, when Wöhler synthesized urea in the laboratory. It is true that natural mixtures tend to be complex, but today their chemistry is no longer regarded as special or unaccountable. We are now able to prepare in the laboratory thousands of compounds which are identical to substances found in nature. The word "synthetic," therefore, should not be taken to imply inferior quality; the terms "natural" and "synthetic" serve only as convenient designations of the origins of substances.

The solvents that are familiar to the conservator are representative of the fundamental types of organic compounds. Because some conservators may have little formal training in chemistry, an attempt will be made in this chapter to develop a general conception of these fundamental compounds. Full justice cannot be done to the subject in so short a space, but perhaps this introduction will encourage the lay reader to explore the subject further.

Organic chemistry is defined as the chemistry of carbon-containing compounds. Carbon is one of the simple substances, designated as *elements*, that combine to make up the world about us. Individual particles of elements are called *atoms*. Two or more atoms may combine chemically to form a *molecule*.

The atoms of the different elements vary in weight. If a weight of 1 is assigned to the lightest atom, that of hydrogen, then the statement that the *atomic weight* of uranium is 238, signifies that an atom of uranium is 238 times as heavy as hydrogen. Oxygen, rather than hydrogen, has, however, become the standard of comparison for atomic weights, with an assigned weight of 16.0000 (giving hydrogen a precise atomic weight of 1.008). The sum of the atomic weights of the atoms in a molecule is its *molecular weight*. That of water, H_2O, is 18.016, for example. *Polymers* are giant molecules, usually made up of carbon, hydrogen, oxygen, and perhaps nitrogen; their molecular weights vary from about 1000 on up to more than 1,000,000.

We find that atoms combine with one another in simple ratios. A hydrogen atom is able to combine with one other atom; oxygen combines with two hydrogen atoms, and carbon with four. Chemists say that the *valence* of hydrogen is one; that of oxygen, two, and that of carbon, four. In drawing the structure of molecules, valence bonds are frequently indicated by a line, as in the following examples of water and of methane:

$$H-O-H \qquad\qquad H-\underset{\underset{H}{|}}{\overset{\overset{H}{|}}{C}}-H$$

The carbon atom has the ability to combine extensively with other carbon atoms. This is a property possessed by no other element to the same degree, although to a limited extent similar compounds are formed by germanium, silicon, and a few others. In the discussion which follows it will be seen that the ability of a carbon atom to combine with another leads to the formation of compounds in wide variety.

HYDROCARBONS

The organic compounds which contain only atoms of carbon and hydrogen in their molecules are called *hydrocarbons*. In the simplest class are the paraffin hydrocarbons. These substances are characteristically inert and show little tendency to combine with many reagents, hence the name, derived from the Latin *parum affinis* (slight affinity). Figure 2-1 presents a number of compounds of this class. Notice that each hydrogen atom has been given one valence bond and that each carbon atom is associated with four valence bonds.

The boiling points of these compounds, also given in Figure 2-1, illustrate a rule of thumb which may provide a useful working concept: the larger the molecule of a given class, the lower the volatility. For example, under ordinary conditions of temperature and pressure, the normal paraffins with from

FUNDAMENTAL ORGANIC COMPOUNDS 9

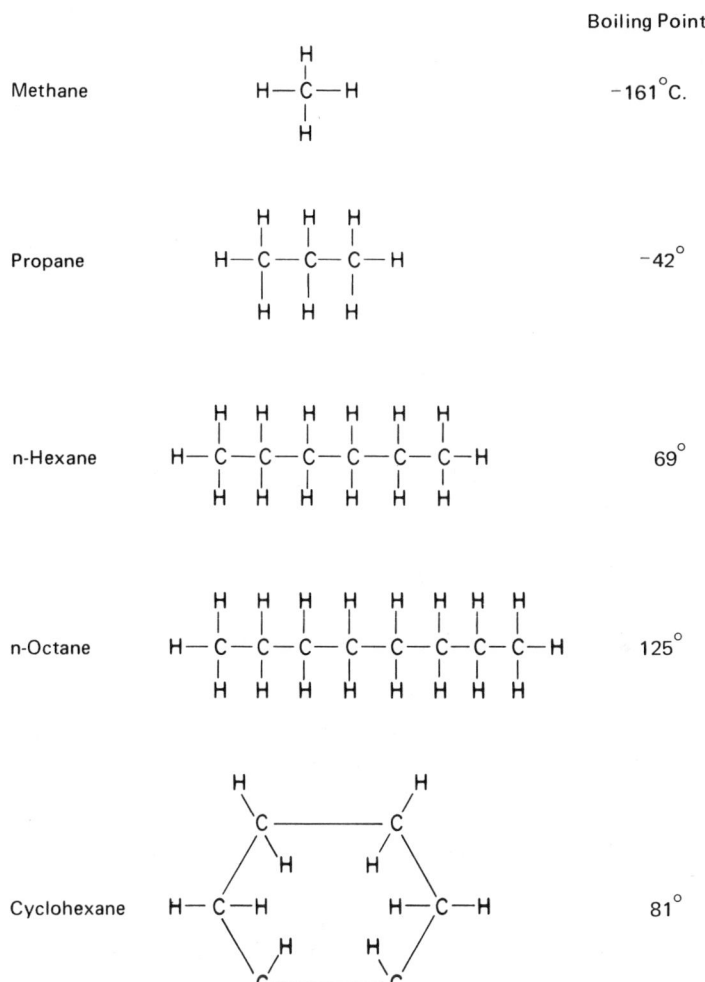

Figure 2-1 Paraffins.

one to four carbon atoms are gases. Those having from five to sixteen carbon atoms in a row are liquids at room temperature and atmospheric pressure. Molecules having more atoms are so large that they form solid substances having the characteristics of the familiar paraffin waxes. Ozokerite, ceresin, or earth wax is composed of paraffin hydrocarbons, as are also the modern "microcrystalline waxes."

When the carbon atoms are arranged in a chain such that each carbon atom is attached to no less than two hydrogen atoms, the compounds are designated as *normal*. Note, however, that in the case of butane, the four carbon atoms can be arranged in two different ways:

```
    H H H H                           H H H
    | | | |                           | | |
H — C—C—C—C — H  n-Butane     H — C—C—C — H  Isobutane
    | | | |                           |   |
    H H H H                           H   H
                                      H—C—H
                                          |
                                          H
```

Two butanes are known. Of the pentanes, three compounds are known; of the hexanes, five. Two compounds that are made up of the same number and kinds of atoms (as shown by the fact that they have the same percentage composition and molecular weight) are termed *isomers*; we say that isobutane is an isomer of butane, that there are five isomers of hexane. This concept is an important one: isomerism is another cause for the vast variety of compounds in organic chemistry; for example, there are 1858 isomers of a hydrocarbon containing fourteen carbon atoms.

Consider next the aromatic family of hydrocarbons, another major classification. The most familiar compounds, benzene, toluene, and xylene, are shown in Figure 2-2. (Carbon atoms have been omitted in these diagrams for convenience.) The reasons why these differ in properties from the paraffin hydrocarbons need not be fully discussed; note, however, that cyclohexane has a composition C_6H_{12}, whereas benzene has the formula C_6H_6: at first glance, six valence bonds seem to be missing. Kekulé suggested that these valences were utilized in the formation of what are called "double" bonds. By placing three double valence bonds around the ring, we are able to hold to the rule that each carbon atom must have four valence bonds. An additional conception of why these are different from the cycloparaffins may be gained by noting, in Figure 2-2, that the hydrogen atoms around the "benzene ring" lie in the same plane as the six carbon atoms, whereas the six carbon atoms in cyclohexane and their hydrogen atoms do not lie in a plane.

The fact that the larger molecules have lower volatility may again be illustrated by the series: benzene, toluene, xylene. Toluene may be designated as methyl benzene and xylene, dimethyl benzene. Obeying the rules of valence, one may devise three xylenes: ortho-, meta-, and para-. These exist and have different properties, such as their boiling points, which are 144.0°, 139.3°

FUNDAMENTAL ORGANIC COMPOUNDS 11

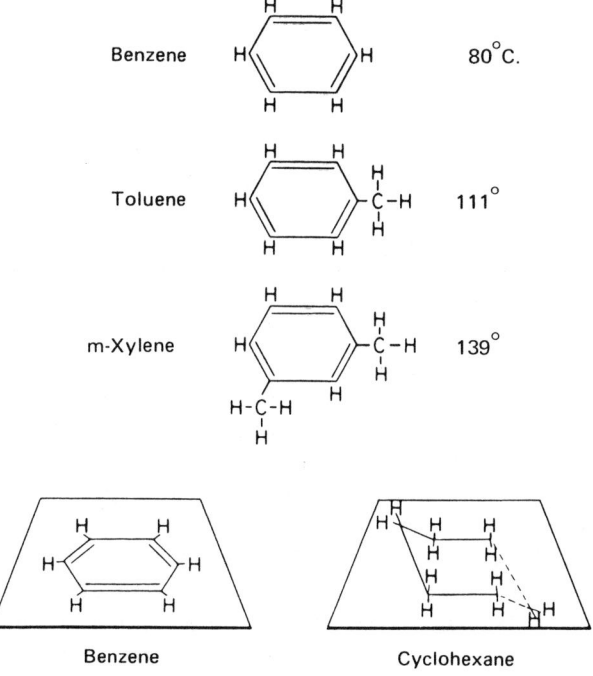

Figure 2-2 Aromatic Hydrocarbons and Cyclohexane.

and 138.5°C., respectively. They differ according to whether the two methyl groups are on carbon atoms that are adjacent, once removed, or on opposite sides of the ring. In a familiar insecticide, para-dichlorobenzene, for example, the chlorine atoms are located on directly opposite carbon atoms in the benzene ring. Giving numbers to the carbon atoms in the ring, this compound may also be called 1,4-dichlorobenzene.

PETROLEUM FRACTIONS

Crude oil is a complex mixture of hydrocarbons along with traces of other substances. Through distillation the hydrocarbons are separated into "fractions" having various boiling ranges. The fractions are still fairly complex mixtures of hydrocarbons, although most of the trace substances containing oxygen, sulfur, and nitrogen are usually removed by various methods of refining. The following list is a classification given by one author for some of the fractions found in commerce.[1]

Fraction	Boiling Range °C.
Petroleum ether	40° to 70°
Gasoline	70° to 90°
Naphtha	80° to 120°
Benzine	120° to 150°
Kerosene	150° to 300°

The composition of petroleum is frequently expressed in the percentages of paraffins, naphthenes (cycloparaffins), and aromatics; the proportions of these compounds vary widely, depending in large measure upon the crude stock.[2,3] An excellent description of the components in petroleum is found in a 90-page booklet by G. W. Waters, *Paint Thinners from Petroleum* and in Reynolds.[4,5] *The Petroleum Thinner Index*, Circular 761, published by the Scientific Section, National Paint, Varnish and Lacquer Association, Inc., Washington, D.C., lists hundreds of petroleum solvents available in America.

The concentration of aromatic compounds present strongly influences the solvent action of petroleum, as does to a lesser degree the ratio of naphthenes to paraffins.[3,5,6] The names "mineral spirits" and "solvent naphtha" are thus not as informative to the chemist as the specification of the boiling range and the ratio of the principal constituents in the petroleum. Although information regarding the three classes of chemical constituents may not always be available, a reasonable specification involves the boiling range and the percentage of aromatic compounds. If the supplier is not able to report the aromatic content, the Kauri-Butanol value of the petroleum (a test described in Chapter 3) may be used as an indication of the aromatic content, as shown in Figure 3-7.[5,7,8,9] The density and refractive index is also useful.[3,9] Tests for aromatic content may also be made with sulfuric acid or by absorption on silica gel.[3,10]

Since the aromatic content profoundly influences the solvent action of petroleum, the conservator may wish to use a non-aromatic petroleum to which known amounts of aromatic compounds may be added in order to obtain greater precision in controlling solvent action; A. P. Laurie, in discussing such *restrainers*, called attention to the usefulness of petroleum fractions free from aromatic hydrocarbons.[11]

The specification of British white spirit relates primarily to flash point and boiling range and does not specify the amount of aromatic compounds present.[12] Two samples tested at Mellon Institute were found to contain from 16 to 20% aromatics. There are petroleum fractions which are poorer solvents than these samples of white spirit, although the solvent may be considered to be a "mild" one.

The names benzene, toluene, and xylene are the internationally accepted designations for aromatic compounds. Aromatic compounds are sometimes designated by other names, such as benzol or toluol. If such terms are used by a supplier there is a chance that a proprietary mixture of hydrocarbons may be involved; one must be alert for this possibility. The word "benzine"– ending "ine"–stands for a mixture of hydrocarbons, usually highly paraffinic.

TURPENTINE

A few remarks are appropriate concerning turpentine although it is being used with less frequency by conservators.[13,14,15] *Gum turpentine* is the steam-volatile portion of the oleoresin that exudes from incisions in the trunks of pine trees. The volatile portion is separated into a turpentine fraction of boiling range 155 to 165°C. The composition of turpentine depends largely upon the species of tree; the principal constituent, however, is α-pinene, shown in Figure 2-3. Other constituents are beta-pinene, monocyclic

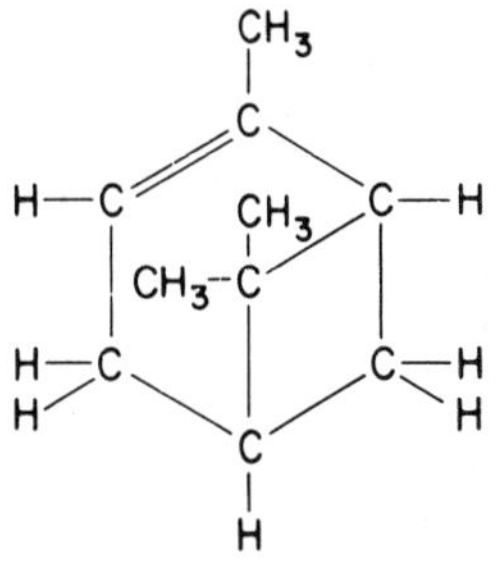

α-Pinene
155–156°C.

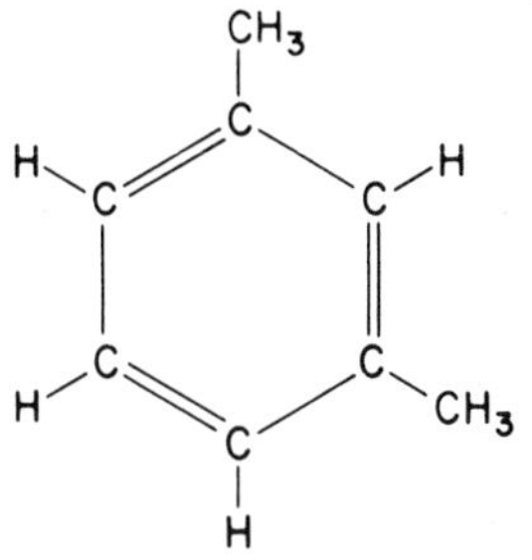

m-Xylene
139°C.

Figure 2-3

terpenes, and terpene alcohols. α-pinene takes up oxygen readily from the air, forming first a peroxide. If moisture is also present, pinol hydrate may be formed as a product of the oxidation and separate as crystals from the turpentine. The other terpenes also easily oxidize and polymerize, causing residues, often of yellow or brown color, to form in turpentine.

The boiling range observed in turpentine illustrates another useful point to bear in mind. Pure substances have sharply defined properties; the "boiling point" of a pure compound, therefore, is a precise temperature, seldom considered to cover a range of more than one degree. On the other hand, a mixture of substances, each of a different boiling point, generally exhibits a considerable boiling range; compare the boiling range of turpentine, a mixture, with that of α-pinene, an organic compound. Recognition of this property of pure substances would be one way of judging whether a commercial solvent given the name "Toluol" is reasonably close to the pure compound, toluene: if the solvent has a broad boiling range, one would suspect that it is a mixture of substances of which the major constituent may be toluene.

UNSATURATED HYDROCARBONS

One double bond is present in alpha-pinene. Compounds which contain one or more double bonds of this character are called *unsaturated* compounds, olefins, or *ethylenic* hydrocarbons. *Acetylenic* hydrocarbons contain a triple bond. These two classes of compounds form another important segment of organic chemistry; they tend to be more reactive than the paraffinic or *saturated* hydrocarbons. The ethylenic compounds are of considerable importance, for they are the basis for the formation of many familiar polymers. It is the presence of ethylenic double bonds in linseed oil that leads to the oxidizing, polymerizing, and hardening properties of linseed oil. It is the presence of this same type of bond in vinyl acetate, methacrylic esters, and styrene which leads to the formation of poly(vinyl acetate), polymethacrylates, and polystyrene.

The double bonds in an aromatic (benzene) ring are in alternate positions, that is, every other carbon-to-carbon bond is double. In the closed ring of six carbon atoms, these alternate double bonds are not as reactive as the usual bonds of ethylenic character (recall that α-pinene easily oxidizes). The bonds in the benzene ring are said to have *aromatic character* and do not have high reactivity.

The classes of compounds discussed thus far involve only hydrogen and carbon atoms. Nevertheless, a vast number of compounds are represented. Important subdivisions of the hydrocarbon classification have been designated: (1) the paraffinic or saturated hydrocarbons; (2) the unsaturated hydrocarbons, including ethylenic hydrocarbons, also called olefins, and the acetylenic hydrocarbons; (3) the cycloparaffins, of which derivatives of cyclopentane and cyclohexane are perhaps the best known; and (4) the aromatic compounds.

DERIVATIVES OF PARAFFIN HYDROCARBONS: ALCOHOLS

Next, let us consider the organic compounds which contain oxygen as well as carbon and hydrogen. If we take a molecule of water, having a formula H_2O, and replace one of the hydrogen atoms with a carbon atom and its

FUNDAMENTAL ORGANIC COMPOUNDS 15

associated hydrogen atoms, we have a member of the family of compounds known as *alcohols*. Several types are illustrated in Figure 2-4.

These compounds may be given the symbol ROH; the R denotes an organic radical. If a hydrogen atom is removed from the paraffin hydrocarbons, groups of atoms remain which have no separate existence but which form combinations with other atoms or groups of atoms. These are called the alkyl

		Boiling Point	Per Cent OH
Water	HOH	100°C.	95
Methanol	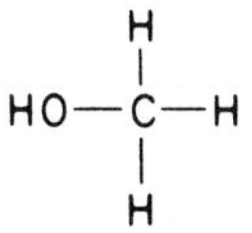 HO—CH$_3$	64.7°	53
Ethanol	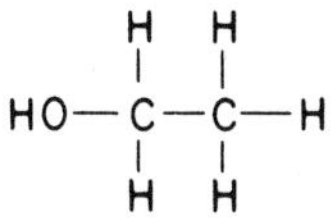 HO—C$_2$H$_5$	78.0°	37
Isopropanol	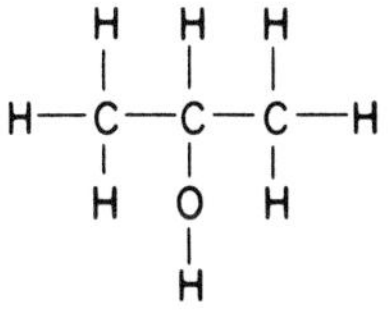 H—C(H$_2$)—C(OH)(H)—C(H$_2$)—H	82.5°	28
Ethyl Cellosolve	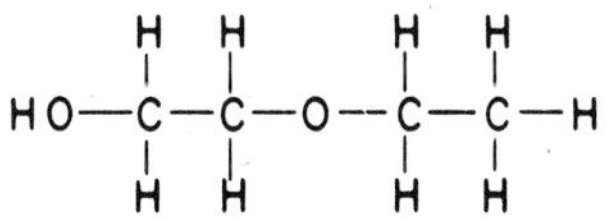 HO—CH$_2$—CH$_2$—O—CH$_2$—CH$_2$—H	135.1°	
Stearyl Alcohol	HO—C$_{18}$H$_{37}$	m.p. 58.5°	6.3

Figure 2-4 Alcohols.

radicals and are characterized by the ending "yl": CH_3 is methyl, C_2H_5 is ethyl, C_3H_7 is propyl, the names being derived from the corresponding hydrocarbons methane, ethane, and propane. Names of the alcohols follow readily: methyl alcohol, ethyl alcohol, propyl alcohol, etc. A system of nomenclature, the so-called Geneva nomenclature agreed upon by a convention of chemists meeting in Geneva in 1892, prescribes the ending "ol" in naming alcohols: methanol, ethanol, propanol.

In Figure 2-4, note again that the larger molecules generally exhibit lower volatility and higher boiling points. Note also that, in the case of the alcohol containing three carbon atoms, two isomers are possible: normal propanol and isopropanol. The latter is the well-known "rubbing alcohol." Methyl alcohol, the simplest alcohol, is the compound formerly derived from the destructive distillation of wood, commonly called wood alcohol; it is poisonous and causes blindness. Ethyl alcohol is the principal constituent of grain alcohol, resulting from the fermentation reaction of yeast on natural sugar.

A fundamental principle of organic chemistry is that wherever the radical OH is attached to a carbon atom of essentially paraffinic character, the OH group has the characteristics of an alcohol. Likewise, the carboxylic acid radical, COOH, when attached to an alkyl radical, imparts to the molecule the characteristics of an organic acid. Certain properties of alcohols share the characteristics of water, owing to the presence of the OH radical. For example, the lower alcohols, methyl and ethyl alcohol, are miscible with water in all proportions. The hydrocarbon radical is responsible for other characteristics; higher members of the series are less soluble in water and the highest, like the hydrocarbons themselves, are not appreciably dissolved by water. Note, in Figure 2-4, that as the number of carbon atoms in the radical increases, the percentage of OH by weight in the molecule decreases; the character which is given to the molecule by the hydroxyl radical becomes less and less apparent. In the case of the 18-carbon-atom stearyl alcohol, many characteristics of paraffin waxes are evident and the alcohol characteristics are diminished. The hydroxyl radical in an alcohol leads to a greater attraction of one molecule for another than in the case of paraffins. The boiling point of methyl alcohol, for example, is higher than that of methane, and that of ethyl alcohol is higher than that of ethane. One can demonstrate that the alcohol radical in stearyl alcohol exerts its influence upon the molecule by comparing the melting point of stearyl alcohol, 58.5°C., to that of normal octadecane, $C_{18}H_{38}$, melting point 28°.

A series of compounds which differ in composition by CH_2, or a multiple thereof, is denoted as a *homologous* series of compounds. The principle that we have just discussed is based on the fact that the chemical and physical properties do not vary suddenly, but gradually, from member to member of a homologous series. Knowledge of the behavior of one member thus enables us to draw conclusions regarding the properties of lower or higher members of the series. However, the change in properties among the first members of the series, from methyl alcohol to ethyl and propyl, vary more sharply than the change in properties among those higher in the homologous series. For this reason, one must not assume that methyl alcohol, which is available with few government restrictions, will do all the same things for the conservator that ethyl alcohol will.

A mixture of 95.5% alcohol and 4.43% water by weight boils at 78.2°C., a temperature slightly lower than the boiling point of *absolute* (100%) ethyl alcohol, 78.3°. Ordinary pure ethanol is usually this constant boiling mixture, designated as 95% alcohol. The absolute, or water-free, alcohol is much more expensive to prepare and is usually not necessary for most conservation needs.

In order to control the distribution of intoxicating beverages, most governments make it legally difficult to secure pure ethyl alcohol, except for research purposes. Readily available is "denatured" alcohol, which is ethyl alcohol rendered unfit for human consumption by the addition of certain compounds from which it cannot be separated by distillation.[16] One formula for "completely denatured" alcohol calls for the addition of kerosene, methyl isobutyl ketone, and acetaldol.[16] The conservator should be aware that such additives may occasionally alter the solvent action of the alcohol.

ACIDS, ESTERS, KETONES, AND ETHERS

Let us now discuss the other major classes of organic compounds containing carbon, hydrogen, and oxygen. Organic acids are designated by the general formula RCOOH. The properties of acids are characterized by the fact that they liberate hydrogen ions in water. The hydrogen which is readily removed as a positively charged ion in these *carboxylic acids* is the hydrogen which is attached to the oxygen in the COOH radical.

Acetic acid, from vinegar, is shown in Figure 2-5. It is the simplest organic acid that contains a paraffinic radical. Formic acid has the formula HCOOH and differs somewhat in properties from the other organic acids, although it may be considered to be the first member of a homologous series. The names of organic acids with which the conservator is perhaps most familiar are the acids derived from fats and oils: palmitic acid, $C_{15}H_{31}COOH$; stearic acid, $C_{17}H_{35}COOH$; oleic acid, $CH_3(CH_2)_7CH=CH(CH_2)_7COOH$; and linoleic $CH_3(CH_2)_4CH=CHCH_2CH=CH(CH_2)_7COOH$. The potassium and sodium salts of such acids, where a sodium ion or a potassium ion replaces the acid hydrogen, are the common fatty-acid soaps.

THE SOLVENTS

		Boiling Point
Acetic Acid	$H-O-\overset{O}{\underset{}{C}}-\overset{H}{\underset{H}{C}}-H$	118°C.
Ethyl Acetate	$H-\overset{H}{\underset{H}{C}}-\overset{H}{\underset{H}{C}}-O-\overset{O}{\underset{}{C}}-\overset{H}{\underset{H}{C}}-H$	77°
Acetone	$H-\overset{H}{\underset{H}{C}}-\overset{O}{\underset{}{C}}-\overset{H}{\underset{H}{C}}-H$	56°
Diethyl Ether	$H-\overset{H}{\underset{H}{C}}-\overset{H}{\underset{H}{C}}-O-\overset{H}{\underset{H}{C}}-\overset{H}{\underset{H}{C}}-H$	34.6°

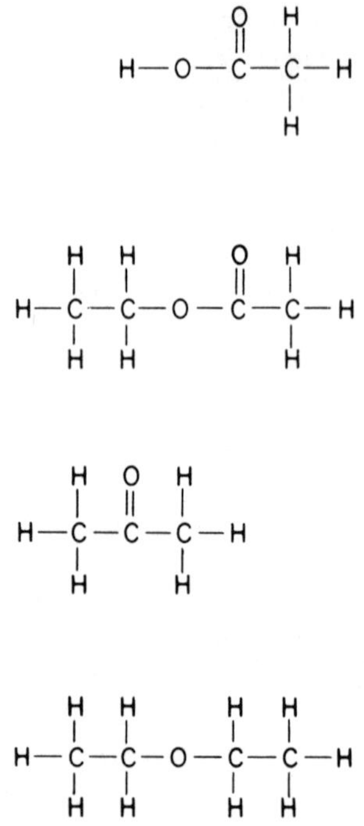

Figure 2-5 Typical Organic Acid, Ester, Ketone, and Ether.

Although the Geneva nomenclature for the carboxylic acids exists, more frequently the common names are used: formic acid, HCOOH; acetic acid, CH_3COOH; propionic acid, C_2H_5COOH; butyric acid, C_3H_7COOH. (Note, in naming these, that butane, with four carbon atoms, forms the basis for naming butyric acid, not the radical C_3H_7 in RCOOH.)

If the hydrogen on the carboxyl group is replaced by an organic radical, a class of compounds with the general formula RCOOR, known as *esters*, is derived. Esters represent an important class of solvents in the industrial formulation of paints and varnishes. The nomenclature is simple; the examples in Figure 2-6 are well worth studying.

Esters are the result of a reaction between an alcohol and an organic acid in which a molecule of water is lost. See Figure 2-7. An important class of condensation polymers are polyesters.

A type of ester of major interest to conservators is found in the natural fats and oils, tri-esters of glycerine. Glycerine is a tri-hydroxy alcohol with the

Methyl Acetate

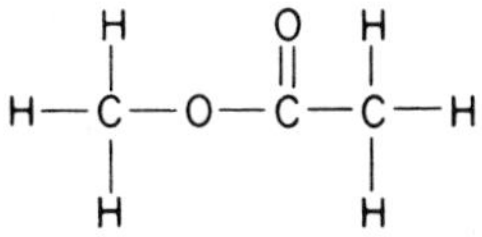

$$\text{H}-\underset{\underset{\text{H}}{|}}{\overset{\overset{\text{H}}{|}}{\text{C}}}-\text{O}-\overset{\overset{\text{O}}{\|}}{\text{C}}-\underset{\underset{\text{H}}{|}}{\overset{\overset{\text{H}}{|}}{\text{C}}}-\text{H}$$

Ethyl Acetate

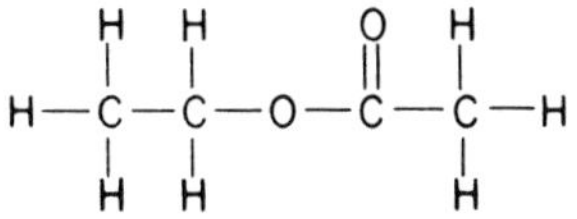

$$\text{H}-\underset{\underset{\text{H}}{|}}{\overset{\overset{\text{H}}{|}}{\text{C}}}-\underset{\underset{\text{H}}{|}}{\overset{\overset{\text{H}}{|}}{\text{C}}}-\text{O}-\overset{\overset{\text{O}}{\|}}{\text{C}}-\underset{\underset{\text{H}}{|}}{\overset{\overset{\text{H}}{|}}{\text{C}}}-\text{H}$$

Methyl n-Butyrate

$$\text{H}-\underset{\underset{\text{H}}{|}}{\overset{\overset{\text{H}}{|}}{\text{C}}}-\text{O}-\overset{\overset{\text{O}}{\|}}{\text{C}}-\underset{\underset{\text{H}}{|}}{\overset{\overset{\text{H}}{|}}{\text{C}}}-\underset{\underset{\text{H}}{|}}{\overset{\overset{\text{H}}{|}}{\text{C}}}-\underset{\underset{\text{H}}{|}}{\overset{\overset{\text{H}}{|}}{\text{C}}}-\text{H}$$

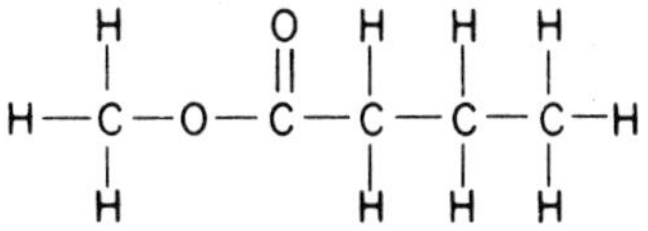

Isopropyl Propionate

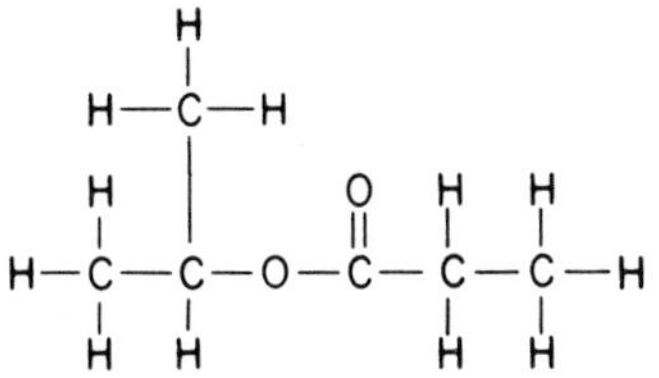

Figure 2-6   Examples of Esters.

following structural formula:

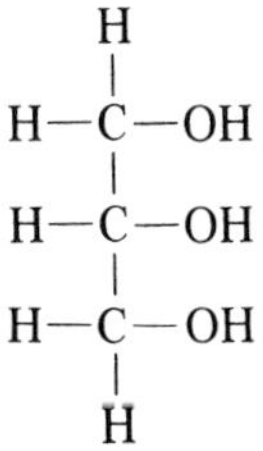

$$\begin{array}{c}\text{H}\\|\\\text{H}-\text{C}-\text{OH}\\|\\\text{H}-\text{C}-\text{OH}\\|\\\text{H}-\text{C}-\text{OH}\\|\\\text{H}\end{array}$$

In linseed and other natural oils not all of the fatty-acid radicals in the tri-ester are alike, a fact that Dr. Stolow will discuss in Chapter 4. Various triglycerides, such as the dilinoleo-linoleneo and the oleo-linoleo-linoleneo, have been obtained from linseed oil.

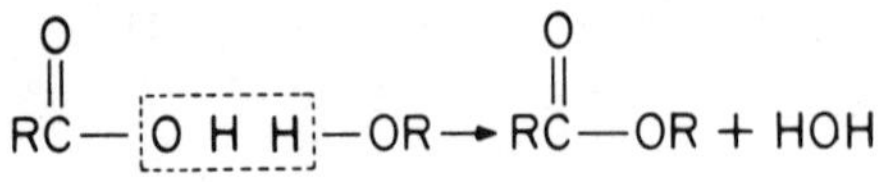

$$\underset{\text{RC}}{\overset{\text{O}}{\|}} - \boxed{\text{O H H}} - \text{OR} \rightarrow \underset{\text{RC}}{\overset{\text{O}}{\|}} - \text{OR} + \text{HOH}$$

Figure 2-7 Typical Esterification Reaction.

Returning again to the formula of water: if both hydrogens are replaced by an organic radical, we obtain compounds of the formula ROR, called *ethers*. The nomenclature is simple: dimethyl ether has two methyl radicals attached to oxygen, and diethyl ether has two ethyl groups. Mixed ethers are also known, such as methyl ethyl ether. The best known ether, the anesthetic, was traditionally diethyl ether; divinyl ether has now come into greater usage in medicine. Note the lowering of the boiling point of the compound that occurs upon substituting both hydrogens in water by an ethyl radical: compare the boiling point of diethyl ether with that of ethyl alcohol and with that of water.

Compounds with the formula RCOR, the structural formula of acetone in Figure 2-5, are known as *ketones*. Along with the esters, ketones form an important class of solvents in modern paint and varnish technology. Again the

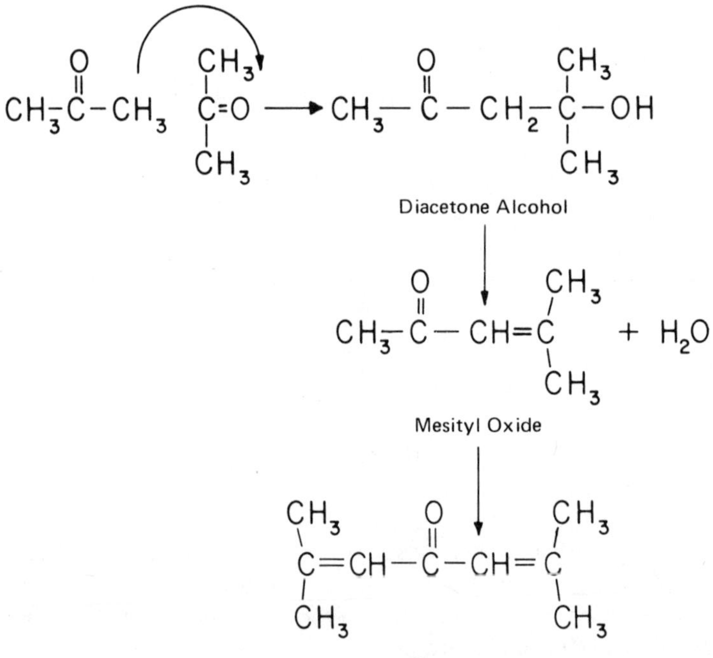

$$CH_3-\overset{O}{\underset{\|}{C}}-CH_3 \quad \underset{CH_3}{\overset{CH_3}{\underset{|}{C=O}}} \longrightarrow CH_3-\overset{O}{\underset{\|}{C}}-CH_2-\underset{CH_3}{\overset{CH_3}{\underset{|}{C}}}-OH$$

Diacetone Alcohol

$$\downarrow$$

$$CH_3-\overset{O}{\underset{\|}{C}}-CH=\underset{CH_3}{\overset{CH_3}{C}} + H_2O$$

Mesityl Oxide

$$\downarrow$$

$$\underset{CH_3}{\overset{CH_3}{C}}=CH-\overset{O}{\underset{\|}{C}}-CH=\underset{CH_3}{\overset{CH_3}{C}}$$

Phorone

Figure 2-8 Diacetone Alcohol and Closely Related Compounds.

FUNDAMENTAL ORGANIC COMPOUNDS

nomenclature is simple: acetone is dimethyl ketone; two of the better-known industrial solvents are methyl ethyl ketone and diisobutyl ketone.

The structure of diacetone alcohol may be of interest. The compound is formed by the combination of two molecules of acetone. This compound is rather reactive and may, by loss of water, form mesityl oxide. Through a similar reaction, the compound phorone may be formed from mesityl oxide, as is shown in Figure 2-8. Diacetone alcohol is frequently yellow in color and is usually distilled in order to render it colorless. Elm[17] has pointed out that phorone is a colored compound; it may be that the formation of phorone-like products in diacetone alcohol leads to the discoloration of this solvent. Cellosolve acetate, a "milder" solvent, has been substituted for diacetone alcohol in poly(vinyl acetate) retouching mediums to avoid the tendency of the varnish to turn yellow in the bottle.[18]

ORGANIC COMPOUNDS CONTAINING HALOGENS AND NITROGEN

The valence of the halogens is one: fluorine, chlorine, bromine, and iodine. From these are derived halogenated hydrocarbons such as chloroform, $CHCl_3$; carbon tetrachloride, CCl_4; and others familiar to the conservator. 1,1,1-trichloroethane (the 1's signify that chlorine atoms are on the first carbon in the chain) is $CCl_3\text{-}CH_3$. The valence of nitrogen is generally three; ammonia gas has the formula NH_3. Figure 2-9 shows a few amines, organic compounds related to ammonia.

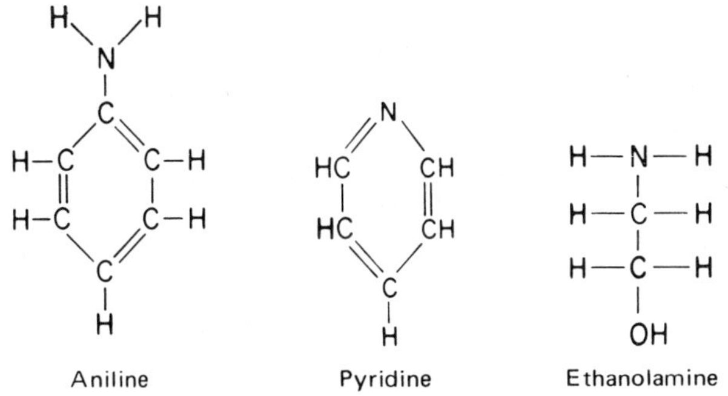

Aniline   Pyridine   Ethanolamine

Figure 2-9 Familiar Organic Amines.

SUMMARY

The major classifications of organic compounds have been discussed. These include: hydrocarbons; alcohols; carboxylic acids; esters; ethers; ketones; and a few compounds containing nitrogen and halogen. An attempt has been made to indicate the system of nomenclature for the simple compounds so that the structure of the common solvents in the museum laboratory can be better understood. Attention has been drawn to a general concept that, within a homologous series of any type of compound, liquid compounds with a greater number of carbon atoms evaporate less readily; or, in other words, have higher boiling points. This rule cannot be readily extended, however, to comparisons between different classes of compounds, as is clearly demonstrated by the example of the changes in boiling point from water to ethyl alcohol to ethyl ether.

Members of a homologous series of compounds tend to keep their identity as alcohols, ketones, acids, esters, etc. This concept is useful to the conservator; for example, if he finds that ethyl acetate dissolves a substance satisfactorily, but that it evaporates inconveniently fast, the first alternative is to try the next higher member of the homologous series, propyl acetate.

In attempting to apply higher members in a homologous series to a problem in which the lower members have proven to be satisfactory, it is important to bear in mind the aspect first pointed out in the discussion of alcohols. There it was shown that as the percentage of OH, the hydroxyl group, decreases, or, to express it differently, as the number of carbons in the homologous series increases, the paraffinic character begins to play a more important role than the hydroxyl character of the molecule. Hence, where the lower members of a homologous series may be very good solvents in a particular situation, the higher members of the series tend to behave more as one would expect of paraffinic hydrocarbons, such as hexane, heptane, or octane. In the aromatic series, benzene, toluene, and xylene, for example, two carbon atoms out of eight in xylene, 25%, are paraffinic in character. We may find that xylene or butyl benzene will not dissolve substances that benzene and toluene are able to dissolve; this is dramatically demonstrated in the case of poly(vinyl acetate) which is soluble in toluene at ordinary temperatures but not in xylene. In the tables that appear in Appendix F it may be seen that as the number of carbon atoms in the alkyl radicals increases solvents tend to have lower solubility parameters, similar to those of the hydrocarbons.

A boiling range within one degree usually indicates a reasonably pure compound. In the classification of petroleum fractions the boiling range is a key specification. Where the percentage composition of the various components of petroleum (paraffins, naphthenes and aromatics) is not known, information on the percentage of aromatic compounds is a convenient indicator of the "solvent power."

REFERENCES AND NOTES

1. I. Mellan, *Industrial Solvents* (New York: Reinhold, 1950).
2. W. von Fischer, "Hydrocarbon Thinners," *Paint and Varnish Technology* (New York: Reinhold, 1948), Chapter 14.

3. J. J. Mattiello (ed.), *Protective and Decorative Coatings* (New York: Wiley, 1941), Vol. 1.
4. G. W. Waters, *Paint Thinners from Petroleum*, 2nd ed. (Cleveland, Ohio: Shell Oil Company, Special Products, 1954).
5. W. W. Reynolds, *Physical Chemistry of Petroleum Solvents* (New York: Reinhold, 1963).
6. H. Burrell, *Interchemical Review* 14 (1955): 3-16; 31-46.
7. R. L. Feller, "Identification and Analysis of Resins and Spirit Varnishes," *Application of Science in Examination of Works of Art* (Boston: Museum of Fine Arts, 1958), pp. 51-76.
8. W. von Fischer, *Paint and Varnish Technology*, p. 231.
9. H. A. Gardner and G. G. Sward, *Physical and Chemical Examination of Paints, Varnishes, Lacquers and Colors* (Bethesda, Md.); 12th ed., 1962, Gardner Laboratory, Inc.
10. B. J. Mair, "Separation and Determination of Aromatic and Monoolefin Hydrocarbons in Mixtures with Paraffins and Naphthalenes by Adsorption," *J. Research Nat. Bur. Standards* 34 (1945): 435-51; see also ASTM Test D936-55 and D1319-56T.
11. A. P. Laurie, "Restrainers and Solvents Used in Cleaning Old Varnish from Pictures," *Technical Studies in the Field of the Fine Arts* 4 (1935-1936): 34-35.
12. C. Marsden and S. Mann, "White Spirit," *Solvent Guide* (New York: Interscience, 1963), p. 553.
13. "Terpene Solvents," in Mattiello, *Protective and Decorative Coatings*, Vol. 1, Chapter 23, p. 537.
14. E. Denninger, "Terpentinöl, im Vakuum rektifiziert und seine Lagerfähigkeit" ("The Stability of Vacuum-distilled Turpentine During Storage"), *Maltechnik* 64 (1958): 1-9 [in German]; *IIC Abstracts*, 2878.
15. B. Mühlethaler and S. Giger, "Prüfung von Turpentinölen mit Hilfe der Dünnschichtchromatographie" ("Examination of Turpentine with the Help of Thin Layer Chromatography"), *Farbe und Lack* 70, No. 2 (1964): 116-20 [in German with English summary]; *IIC Abstracts*, 4689.
16. N. A. Lange, *Handbook of Chemistry* (Sandusky, Ohio: Handbook Publishers, Inc., 1956), see Formulas for Denatured Alcohols, p. 1779.
17. A. C. Elm, "Deterioration of Dried Oil Films," *Ind. Eng. Chem.* 41 (1949): 319-24.
18. R. L. Feller, "Research on Durable Thermoplastic Polymers for the Conservation of Works of Art," *Atti della XLIX Riunione SIPS, Siena*, 23-27 September 1967 (Rome, 1968): 1099-1110.

3
VOLATILE SOLVENTS

Previous remarks have reviewed organic compounds from the point of view of their chemical structure, without particular regard to their application as solvents. The nature of solvents and solvent action will now be considered.

SOLUTION

A solution is a homogeneous mixture of the molecules of one material with those of another. It is sometimes designated as a molecular dispersion to emphasize the high degree of dispersion. The molecules may be large, as in the case of polymers, or the units may be aggregates of molecules, but in a true solution the dispersed particles, however constituted, are expected to be less than one nanometer (1 mμ) in size. One may gain some idea of this dimension when it is pointed out that the usual electron microscope today cannot resolve dimensions much smaller than 10 Å, 10^{-7} cm., or 1 nm. The atoms which constitute a molecule are about 1 to 2 Å in diameter.

A solution is distinguished from a colloidal dispersion, in which the dispersed particles range from 1 nm. to about $1/2\,\mu$ in size, by the fact that colloidal particles are sufficiently large that they reflect light. In such dispersions, tiny specks of light can be seen with the special illumination of the Zsigmondy ultramicroscope, although the individual particles cannot be seen. Because of the scattering of light, a foggy cone is visible when a beam of light is passed through a colloidal dispersion (Faraday-Tyndall effect).

When the particles of a dispersion are larger than the colloidal range, the dispersed material may be seen with an ordinary microscope (particles above 0.3 to 0.4 μ). The dispersed material in coarse dispersions or suspensions frequently settles out upon standing.

From these considerations we see that a solution is a limiting case of high dispersion of one material in another. While the conservator may be able to disperse many materials by the action of organic liquids, a true solution is not always obtained; far more frequently a colloidal or a coarse dispersion is made.[1]

VOLATILE SOLVENTS   25

A solvent is simply a material that is capable of dissolving another, that is, capable of dispersing another substance into the submicroscopic molecular dimensions described above. For convenience, the component present in excess is usually designated as the *solvent*; the other, the *solute*. The formation of a solution is not limited to the liquid state, of course, but occurs in the solid and gaseous states as well. The solvents of primary interest in the formulation of solvent-type varnishes are organic compounds that are volatile liquids, often called *volatile solvents*.

The subject of volatile organic solvents may be organized in many ways; so many empirical tests and measurements have been developed that the beginner may be confused by trivialities and lose sight of the essentials. The introductory chapters in Scheflan and Jacobs, however, concentrate very well on the fundamental aspects of the subject.[2] Essentially, these involve classification of solvents according to (a) chemical type and to some indication of (b) volatility and (c) solvent action. Aspects of safety to the user might be used as a basis for a fourth mode of classification.

CHEMICAL TYPE

Classification according to chemical type is fundamental to the chemist. For this reason the subject has been considered separately and at length in Chapter 2. Conservators should attempt to use the purest and best-defined solvent that can conveniently be obtained. "Analytical" or "reagent" grades of compounds are of high quality. "Technical" grades are usually mixtures (indicated by a boiling range of more than two degrees).

By being able to write the formula of the simple compounds taken from the shelf, we "know" in a sense what the substance is; the chemical nomenclature identifies the solvent. Unfortunately, knowing the chemical compound may reveal little directly about its volatility or solvent action.

VOLATILITY

In attempting to classify solvents according to rates of evaporation we are hampered by variations in the behavior of the solvents under different circumstances. Evaporation rates are determined conveniently with an open dish or a strip of saturated filter paper. However, the relative rates measured in this manner frequently are not comparable to the evaporation from coatings, in which the volatility of solvents is influenced by the non-volatile vehicle. Bogin considered this problem some years ago, but more recently, Hansen has published an extensive analysis of the phenomenon.[3,4] Reynolds has also discussed many of the practical problems of measuring and interpreting evaporating rates.[5] Appendix F lists evaporation rates of common solvents relative to ether.

It is well known that the boiling points and evaporation rates are not necessarily proportionate. Bogin suggests, however, that the product of the vapor pressure times the molecular weight of the solvent gives a fair approximation of the experimentally determined *evaporation rate*. Evaporation rates are usually expressed as relative values, as in Table 3-1.

TABLE 3-1. RELATIVE EVAPORATING TIME OF SOLVENTS[5]

(See Appendix F for additional values)

SOLVENT	RELATIVE EVAPORATING TIME
Diethyl ether	1
Acetone	2.1
Benzene	3.0
Toluene	6.1
Methyl alcohol	6.3
Ethyl alcohol	8.3
Xylene	13.5
Turpentine	25.5
Methyl cellosolve	34.5
Diacetone alcohol	147

If evaporation rate data are available, they can be useful. In special research problems, it may be necessary to determine evaporation rates under the conditions of the particular problem at hand. Other than this, one must be content with the boiling point or boiling range as a fundamental specification regarding solvents; boiling characteristics remain the most generally available data related to volatility.

SOLVENT ACTION

Conservators conveniently speak of "strong" and "weak" or "mild" solvents. The meaning is sensed, but seldom defined precisely. The "strength" of a solvent, its power to dissolve, is not an intrinsic property; it is relative to the material to be dissolved. Two aspects of the matter, (a) the rate of attack by a solvent, and (b) the amount of solute that can be dissolved if equilibrium is reached, have not often been considered separately in the designations "strong" or "mild."

The question of whether two substances will form a solution is one of whether an intimate mixture of the two will produce a system (solution) which has less energy than that present under the starting conditions, with the two substances separate. The energy concerned is that which the physical chemist calls the *free energy*. Whether the energy after mixing will be less than that under the starting conditions depends upon the net free energy required, or liberated, in separating the molecules of each material and that required, or liberated, when they come together in homogeneous mixture. The net result is dependent upon: (a) the various forces of attraction between molecules; (b) the randomness of the molecules; and (c) the temperature. We know that raising the temperature usually aids solution (but not invariably); the other two factors have been much more difficult to organize.

Commencing in the 1950s, interest has developed rapidly in the concept of the *cohesive energy density* (CED) or *solubility parameter* (δ) of solvents.[6,7,8,9] Burrell drew special attention to these ideas, which had been developed earlier by Hildebrand, and made the subject more convenient for the coatings chemist to use. Dr. Stolow will discuss the application of solubility parameters to the conservator's problem in the next chapter. Briefly, the nature of the concept is as follows: Numbers may be assigned to solvents, based on their energy of vaporization per cubic centimeter, the cohesive energy density (CED). These numbers are called the solubility parameters (δ) and are equal to the square root of the CED. (The heat of vaporization is a measure of the energy mentioned above, the energy required to separate the molecules.) An amorphous material is ordinarily dissolved (or else swelled considerably) by solvents which have the same solubility parameter as that of the amorphous material. If the parameter of the solvent is much greater or much less than that of the material to be dissolved, solution may not take place or, if there is swelling, it will be appreciably less. In other words, if one of the substances has a much greater solubility parameter than the other, its molecules will tend to cling to themselves more than they will tend to mix with those of the other substance.

It was soon realized that solubility parameter alone, a single number, would not explain solubility behavior with complete success. When a table was added as an appendix to the first edition (1959), the solvents were listed with respect to hydrogen bonding tendency as well. At the time, solubility parameters of polymers and resins were often determined by the range of δ values of the solvents that would dissolve them, three classes of solvents, those of low, medium, and high hydrogen bonding tendency, being established. Values of the mean δ for various polymers and the range of δ values in each of these classes of solvent are shown in Table 3-2. Unfortunately, this added refinement still did not provide a precise system and did not sufficiently "explain" what was taking place.

Beginning about 1965, there have been several developments that have led to further improvements in the original concept.[10] Gardon developed more fully the numbers (or *parameters*) to be assigned for solvent polarity and hydrogen bonding.[11] Crowley, Teague, and Lowe later made a major con-

THE SOLVENTS

TABLE 3-2. THE RANGE OF SOLUBILITY PARAMETERS OF POLYMERS IN VARIOUS SOLVENTS (GARDON)[11]

POLYMER	HYDROGEN BONDING TENDENCY OF SOLVENT		
	POOR	MODERATE	STRONG
Isobutyl methacrylate..	9.0 ± 0.8	8.2 ± 0.8	10.5 ± 0.5
n-Butyl methacrylate ..	9.6 ± 1.4	8.7 ± 1.3	10.5 ± 1.0
Vinyl acetate.......	10.7 ± 2.0	11.6 ± 3.1	14.5 —
Nylon 8..........	—	—	12.7 ± 1.9
Medium oil alkyd	9.0 ± 2.0	— —	9.7 ± 2.3

tribution by suggesting that solubility be considered in terms of three properties: solubility parameter, hydrogen bonding, and dipole moment; they prepared tables that list these properties for common solvents.[12] Figure 3-1, taken from their publication, indicates more clearly than before why some

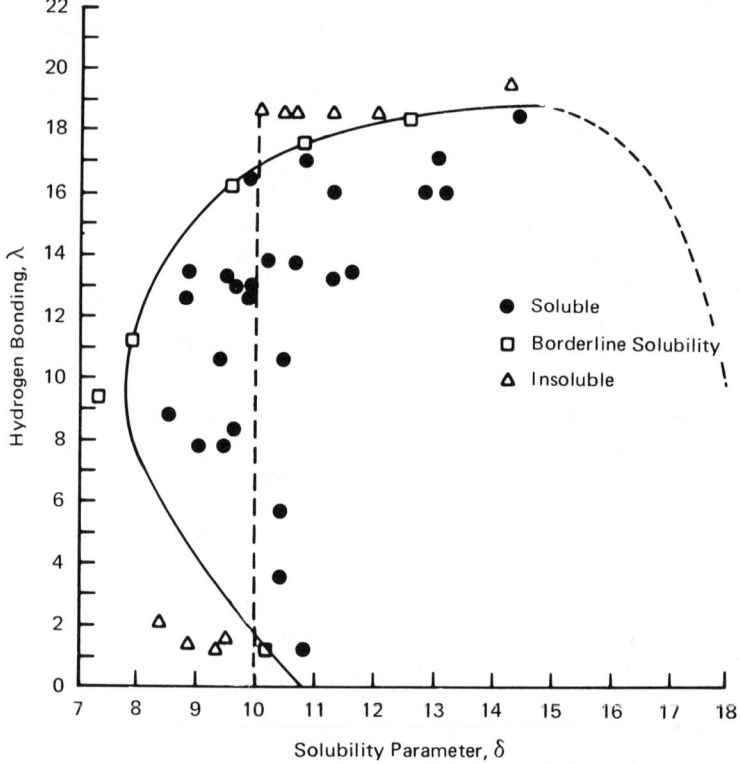

Figure 3-1 Cross-section of Cellulose Nitrate Solubility Plot Through Dipole Moment $\mu = 1.7$ (Crowley, Teague, Lowe).

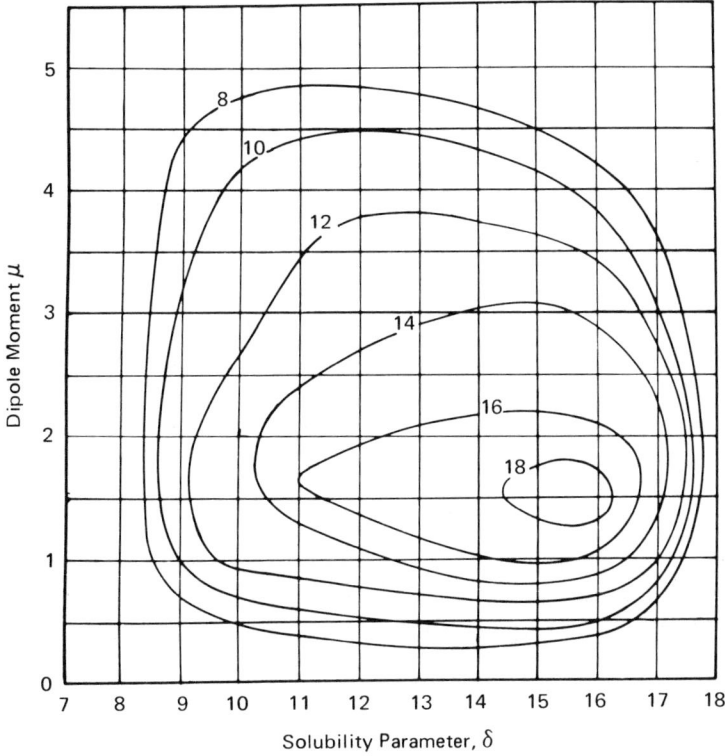

Figure 3-2 Solubility Map of Cellulose Nitrate, Hydrogen Bonding (λ) = 8 to 18 (Crowley, Teague, Lowe).

solvents fail to dissolve the polymer even though they have similar solubility parameters. Combining these data with those of dipole moment, Figure 3-2 summarizes the distribution of the data in the third dimension (the plane perpendicular to the plane of the paper in Figure 3-1). Soon after this work, Hansen developed the same ideas more fully, expressing the original solubility parameter now in terms of three factors based on dispersion, dipole, and hydrogen bonding forces thus:

$$\delta = \sqrt{\delta_d^2 + \delta_p^2 + \delta_h^2}.$$

Hansen, also, gave tables of these values for the common solvents.[13] Figure 3-3 shows a plot of Hansen's data for the solubility of poly(methyl methacrylate); it can immediately be seen that his procedure gives a more precise picture of the "range" of solvent properties which will dissolve a given polymer (his radius R_A). He has also successfully applied these concepts to explain the wetting of pigments, solubility of dyes, and compatibility of

30   THE SOLVENTS

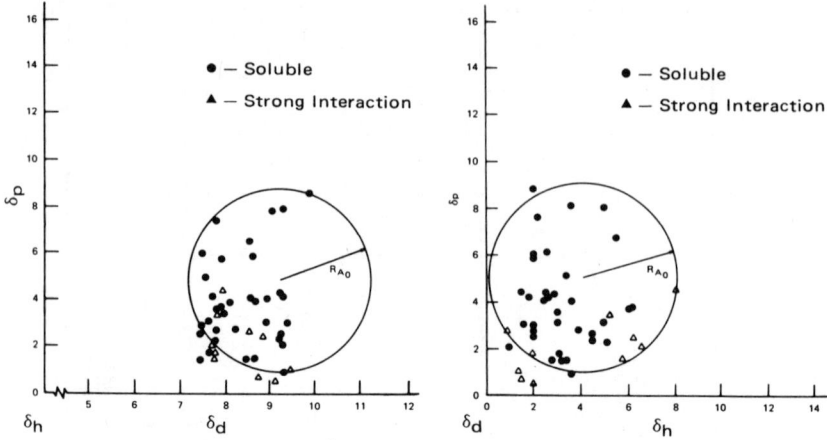

Figure 3-3   The solubility of poly(methyl methacrylate) [Hansen].

emulsifiers and plasticizers. Because the usual three-dimensional plots are somewhat inconvenient to handle, Teas later suggested that triangular graph paper could be used to express these relations, if three functions of the terms of the following type were calculated:

$$f_h = 100\delta_h / \delta_d + \delta_p + \delta_h$$

Figures 3-4 and 3-5 are from his publication.[14]

What are these additional binding forces that must be considered? A hydrogen bond is a force of molecular association in which a hydrogen atom from one atomic combination can attach itself weakly to an atom other than the one to which it is bound through its primary chemical (covalent) bond. Fluorine, oxygen, and nitrogen atoms have the ability to form this weak attachment with hydrogen. Alcohols, for example, whose molecular formulae are usually written ROH, can form associations of molecules having the structure:

$$\underset{H}{\overset{R}{\diagdown}} O \cdots H - \underset{R}{\overset{}{\diagdown}} O$$

Water does the same,

$$\underset{H}{\overset{H}{\diagdown}} O \cdots H - \underset{H}{\overset{}{\diagdown}} O$$

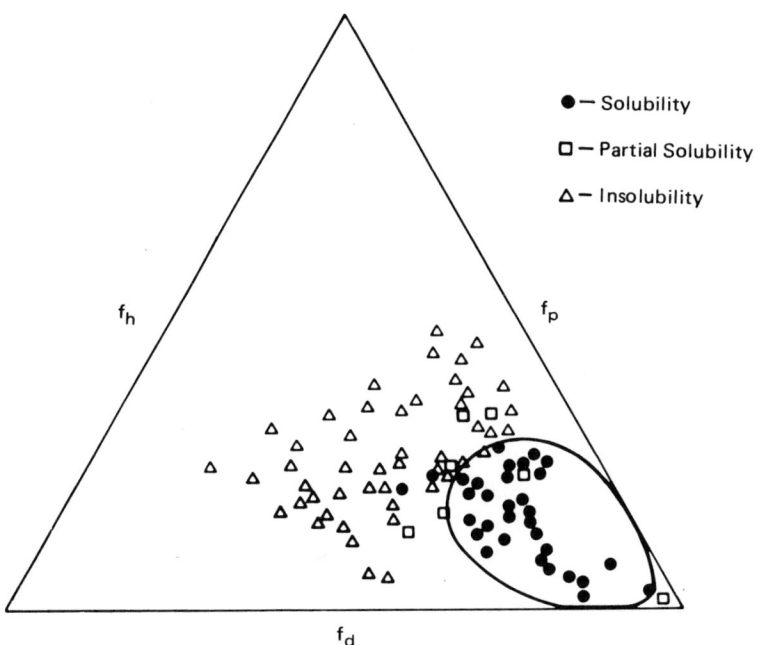

Figure 3-4   Solubility Plot for Piccopale ® 110 Resin (Teas).

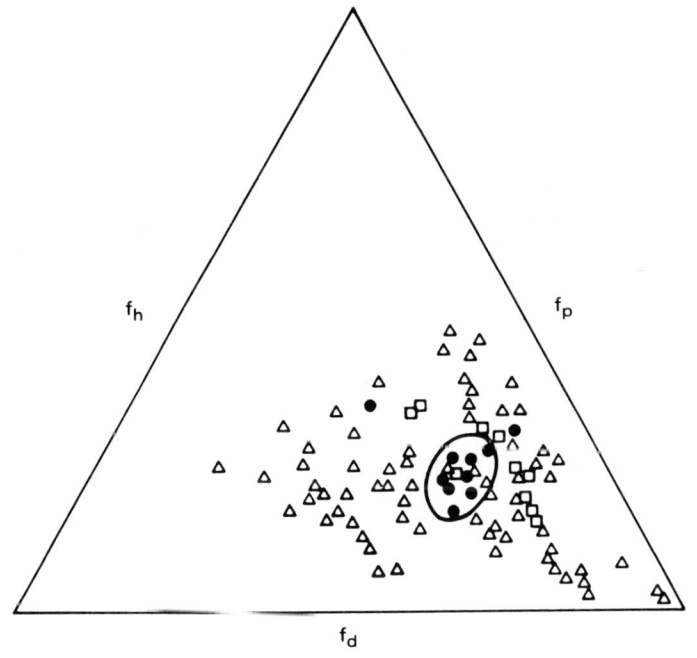

Figure 3-5   Solubility Plot for Vipla KR Poly(vinyl chloride) Resin (Teas).

The atomic weight of acetic acid in benzene is almost twice that expected; the explanation given is that two molecules are associated through hydrogen bonding as follows:

$$CH_3-C{\overset{O \cdots H-O}{\underset{O-H \cdots O}{}}}C-CH_3$$

The hydrogen atom is in effect shared between two molecules in this process; the strength of the weak electrostatic "hydrogen bond" is only about 5 to 6 kcal./mole, whereas the primary chemical bonds in the molecules are from six to twenty times greater. Hydrogen bonding nevertheless contributes important forces of attraction in proteins and cellulose, in the adhesive action of animal glue, and in the structure of ice. Burrell and others have classified solvents from the standpoint of the strength of their hydrogen bonding tendency:

1. Strongly hydrogen-bonded: alcohols, carboxylic acids, pyridine, and water.
2. Moderately hydrogen-bonded: ketones, esters, ethers, and aniline.
3. Poorly hydrogen-bonded: hydrocarbons, halogen compounds, nitro compounds, and nitriles.

Dipolar forces arise because molecules are usually not perfectly symmetrical in shape and in the distribution of their electrons. Hence, they may have a slightly greater negative or positive charge in one part of the molecule than in another. Symmetrical molecules, such as methane, CH_4, and benzene, C_6H_6, do not exhibit this phenomenon. We are familiar with the fact that magnets have an attractive force and readily align themselves with respect to their neighbors. Dipolar (or simply "polar") molecules exhibit similar behavior, owing to the slight separation of electrical charge density that they possess. (The "dipole moment" of a molecule is equal to the product of the charge multiplied by the distance that separates the centers of the positive and negative charge.)

When a polymer dissolves in a solvent, forces between the solvent molecules (SS) and the polymer molecules (PP) must be broken. New forces established between solvent and polymer (SP) can provide the major energy change assisting solution. It may take considerable energy to cause the molecules of polar solvents to separate, but this effect can be offset by a high degree of attraction between solvent and polymer, perhaps through the formation of hydrogen bonds.

A polymer is most likely to dissolve if the dispersion forces (represented by the usual solubility parameter numbers), the dipole forces, and the hydrogen bonding forces of both polymer and solvent are similar. To quote Gardon: "Generally, a liquid will dissolve a polymer if its solubility parameter is within 1 to 2 (cal./cc.)$^{0.5}$ (1 to 2 $\sqrt{\text{cal./cc.}}$) units of that of the polymer.[15] The maximum allowable difference between solvent and polymer solubility parameters is generally increased if the polymer and solvent molecular weights are low, if the polarities are matched and if there is hydrogen bond formation between polymer and solvent.... Polymer solutions in the best solvents will tolerate the addition of largest amounts of nonsolvents without causing precipitation.... Sometimes in industrial practice it is important to find a polymer which is resistant to a given liquid. To achieve this, the differences in solubility parameters and in polarities have to be maximized."

Although the new developments in the solubility parameter concept are still rather complicated for convenient use, we wish to point out the direction that the theories are taking. Through these new developments a better understanding of the perplexing problems of solubility is being achieved.[10]

Solubility parameters relate primarily to the amount of solvent which would be imbibed at equilibrium, that is, if one allowed resin and solvent to remain in contact for a sufficient length of time. The question of the rate of attack by the solvent is another matter. Dr. Stolow has also given attention to this aspect of the problem and discusses it in the next chapter. Prior to his investigations, one could find indications in the literature that smaller molecules would diffuse through a film more rapidly than those of larger volume.[16] The data of Graham (his Table I) tended to bear this out, particularly in the case of the alcohols and in the pair of ketones, acetone and cyclohexanone.[17] These findings imply, as Dr. Stolow points out, that fast-evaporating solvents may not always offer the least risk of solvent action.

So much for the theoretical basis of solvent action; Dr. Stolow discusses in greater detail the practical application of solubility parameters. There are also a number of empirical methods for estimating the relative tendency of substances to form solutions, and mention should be made about some of these. One of the best known techniques is to determine the temperature or composition necessary to obtain a clear mixture as opposed to a cloudy one (the latter condition indicates "dissolution"). A test of this type, developed by the National Gallery of Art Research Project, determines the *solubility grade*, a convenient measure of the relative tendencies for polymers to dissolve in hydrocarbon solvents. All of the methacrylate polymers of interest are soluble in toluene, and only one is soluble in the pure paraffin hydrocarbon, normal dodecane; mixtures of these two solvents thus can be used to measure solubility grade.

Solubility grade is defined by the Research Project staff as "the minimum weight per cent of toluene in normal dodecane required to give a homogeneous, single-phased solution at 25°C. and at a polymer concentration of 10% by weight, using a polymer whose viscosity grade is 20 centipoises." Ten per cent solutions of resin in the appropriate solvent are cooled until they become cloudy. The temperature at which this occurs is called the cloud point.[18]

A plot of cloud point temperature versus per cent toluene was used to determine the composition of solvent at a cloud point of 25°C. as shown in Figure 3-6.[19] The solubility grade of a number of polymers is given in Table 3-3. Those at the bottom of the list are soluble in solvents which are more paraffinic in character than those above.

TABLE 3-3. SOLUBILITY GRADES OF POLYMERS

(Weight Per cent Toluene in n-Dodecane Necessary to Give Clear Solution at 25°C.)

Polymer	% Toluene
Poly(vinyl acetate)	89
Poly(methyl methacrylate)	87
Acryloid ® B-72 (Rohm and Haas)	80
Poly(ethyl methacrylate)	65
Poly(propyl methacrylate)	45
Poly(n-butyl methacrylate)	25
Poly(isobutyl methacrylate)	23
Acryloid ® B-67 (Rohm and Haas)	18
Poly(isoamyl methacrylate)	8
Resin AW-2	4 ± 4

A familiar cloud-point test is the Kauri-Butanol Test used to classify petroleum fractions.[20] In this test a solution of 100 g. of Kauri gum and 500 g. of n-butyl alcohol is prepared. A fixed amount of this solution at 25°C. is diluted with the solvent under test until the mixture becomes cloudy; the volume of solvent added is a measure of the K-B solvency; very little of a "poor solvent" can be added to the stock solution before it becomes cloudy, i.e., lower K-B values indicate lower solvent power.[5,21,22] Figure 3-7 shows the relationship of K-B number to aromatic content in petroleum fractions.[19] Table 3-4 lists K-B numbers of various solvents that may be of interest. Burrell (1957) and Reynolds have given an equation which relates K-B number to solubility parameter, $\delta = 6.9 + 0.02 \times (K\text{-}B)$, but Reynolds has pointed out that below a K-B number of 35 the solubility parameter is not particularly useful in judging the solvent power of branched and straight chain paraffins.[5]

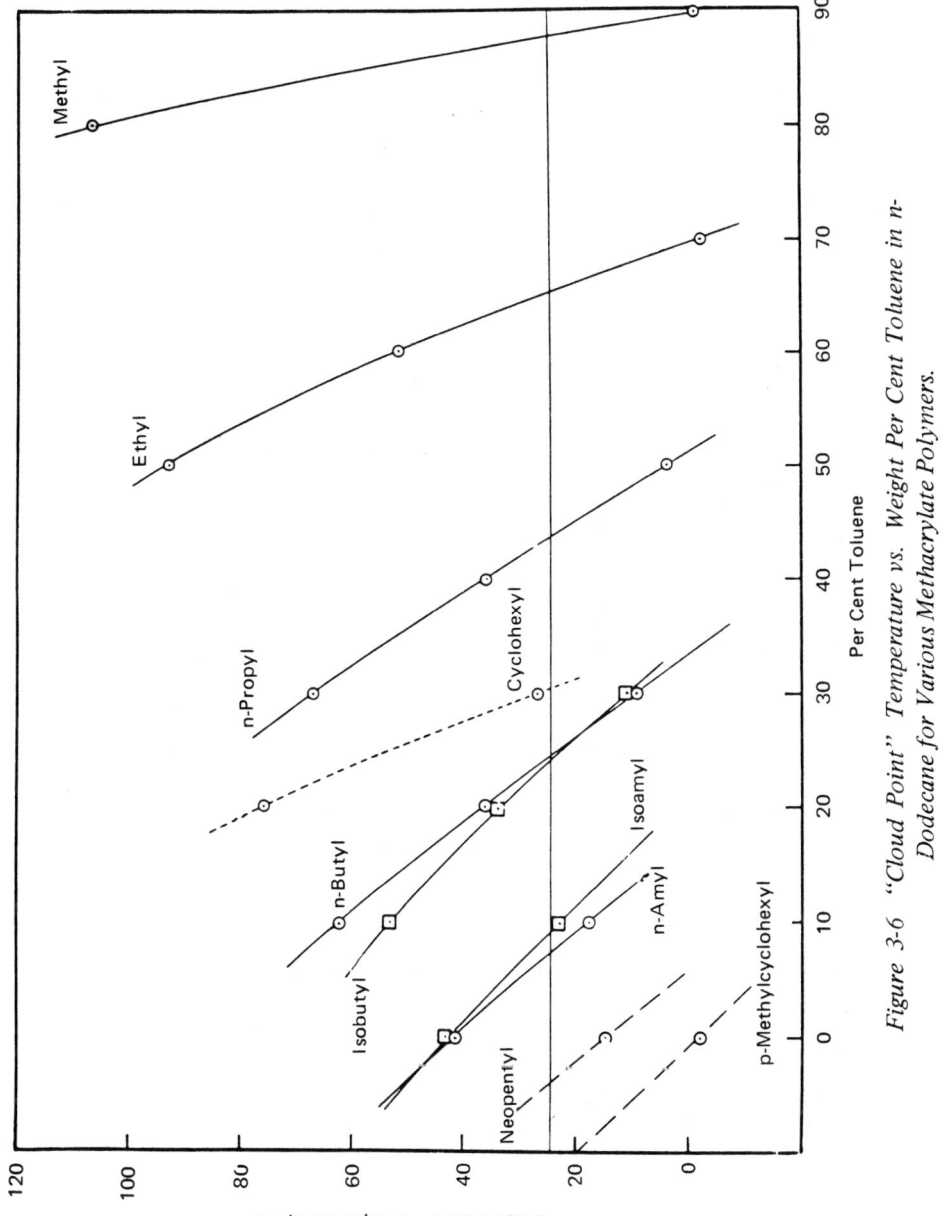

Figure 3-6 "Cloud Point" Temperature vs. Weight Per Cent Toluene in n-Dodecane for Various Methacrylate Polymers.

36   THE SOLVENTS

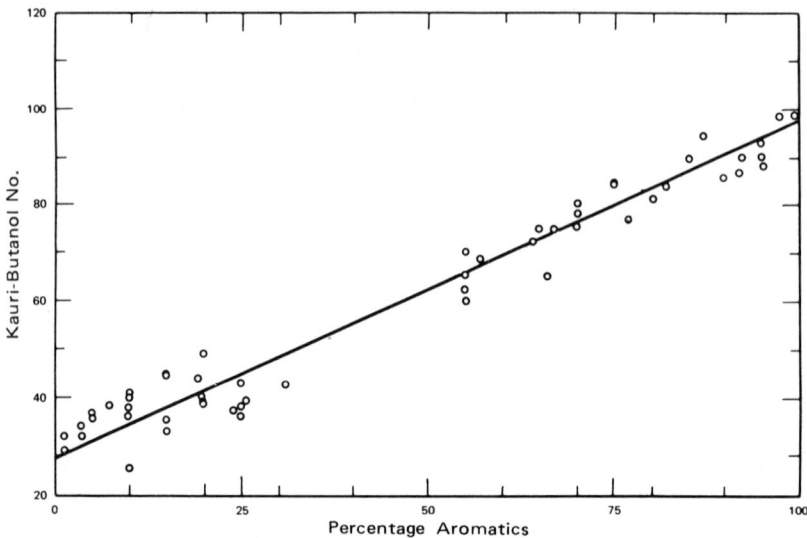

Figure 3-7   A Plot of Data Taken from the Petroleum Thinner Index.

One way of reducing the solvent action of aromatic hydrocarbons is by employing alkyl-substituted derivatives of benzene rather than mixtures of paraffin and aromatic hydrocarbons. According to the concept of Waters, substituted benzenes are considered to be only partially aromatic.[25] Thus, for example, he would consider hexyl benzene to have 50% aromatic character, half the number of carbon atoms being paraffinic. In selecting solvents on this basis, one quickly discovers that volatility and aromatic character are not independent and that the solvents range generally greater than 46% aromatic

TABLE 3-4.   KAURI-BUTANOL NUMBERS OF VARIOUS SOLVENTS AT 77°F,[23,24,25] (25°C.)

SOLVENT	K-B NO.
Aliphatic petroleum naphthas	25–50
n-Heptane	28
"Freon" TF, $CCl_2F-CClF_2$ (du Pont)	31
Turpentine	55–58
Xylene	88
Toluene	107
Carbon tetrachloride	114
Trichlorethylene	130
Chloroform	208

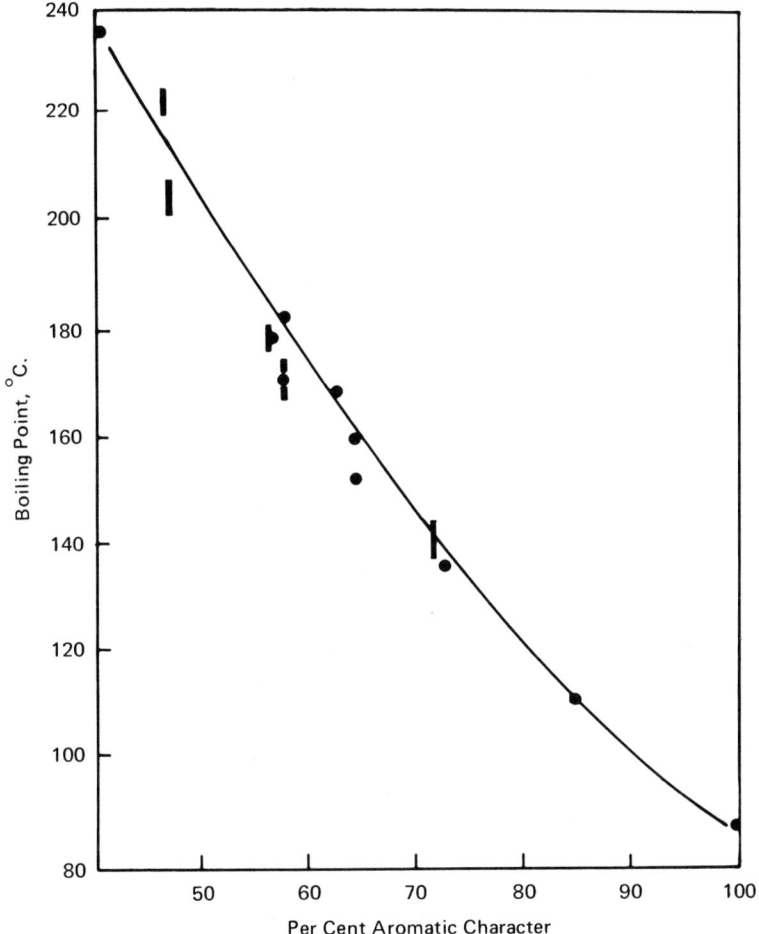

Figure 3-8 Boiling Point vs. Per Cent Aromatic Character: Alkyl Benzenes.

character. Nevertheless, within this range Waters' concept can be useful in selecting a series of solvents of controlled variation in solvent action.

The accompanying Figure 3-8 illustrates the variation of boiling point in this series of aromatic compounds, and Table 3-5 lists their boiling points and approximate cost per kilogram. Certain compounds are more difficult to prepare than others, and because of their high cost, are not listed; however, the reasonably priced solvents in the series may be useful in the conservation laboratory. Diethyl benzene has been suggested as a pure, well-defined solvent for Acryloid ® B-72 when slower evaporation is needed.[26] (The per cent

TABLE 3-5. ALKYLBENZENES

COMPOUND	PER CENT AROMATIC CHARACTER	BOILING POINT °C.	APPROX. COST PER kg.
Benzene	100	87	$ 2.00
Toluene	83.7	110.8	2.00
Ethyl benzene	72.6	136.2	2.95
Xylene	71.7	138–144	1.70
Cumene (isopropyl benzene)	64.2	152.5	8.75
Trimethyl benzene (mesitylene)	62.5	163–166	5.60
Sec-butyl benzene	57.5	173–175	14.00
Diethyl benzene	56.7	175–181	3.00
Di-isopropyl benzene	46.8	202–209	2.90
Triethyl benzene	46.3	220–225	17.70

aromatic character in Table 3-5 is calculated on the basis of the atomic weights of the six carbon atoms of the benzene ring, and their associated hydrogen atoms compared to the molecular weight of the molecule. Waters calculated his percentage on the basis of carbon atoms only.)

SAFETY

The harmful effects of breathing vapors of various solvents should be given careful consideration by the conservator.[27,28] The toxicity of various solvents and their physiological effects may be found in such books as T. H. Durrans' *Solvents*, and in C. Marsden's *Solvents Manual*, as well as information regarding fire hazard.[29,30] (Information on the fire hazard of common solvents is given in Appendix F.) The museum laboratory should have such a handbook among its reference manuals.

It is well to have abundant ventilation at all times when using solvents. At the least, an electric fan can be so arranged to blow the vapors away from the conservator as he works on a particular area of a painting. Lately, portable ventilating hoods have come into use. Covered containers should be used to store waste cotton swabs which contain solvent, both for health and fire safety. Because the physiological effects of solvents depend both on the length of time one is exposed to the vapors and upon the concentration of vapors in the air that is breathed, a closed container for storage of waste swabs and a fan that blows the vapors away from the face of the operator will both aid appreciably in the reducing of vapor concentration.

There is an important aspect in the evaluation of solvent toxicity which may be overlooked. Whereas a person may not be able to tolerate a very high concentration of the vapors of a given solvent, often it can be used with relative safety if its vapor pressure is low. The concentration of vapors in the air owing to the evaporation of diacetone alcohol is lower than the concentration of vapors from chloroform, for example. Similarly, the tolerance of the concentration of mercury vapor is extremely low and yet mercury, "quicksilver," has a very low vapor pressure. An intolerable concentration of its vapor is seldom experienced in cases of brief exposure to an open vessel of mercury.

TABLE 3-6. CALCULATION OF A SAFETY NUMBER FOR COMMON SOLVENTS

	EVAPORATION TIME (ETHER = 1)	MAXIMUM ALLOWABLE CONCENTRATION IN PPM*	SAFETY NUMBER
Benzene	3	35	105
Toluene	6.1	~150	915
Xylene	13.5	~150	2025
Turpentine	25	100	2500
Methyl alcohol	6.3	200	1260
Ethyl alcohol	8.3	1000	8300
Diacetone alcohol	147.1	50	7350
Methyl ethyl ketone	6.3	200	1260
Acetone	2	400	800
Chloroform	~2	50	100
Carbon tetrachloride	~5	75	375

*Limits of concentration to be tolerated under continuous working conditions, from *Handbook of Organic Industrial Solvents*, Chicago, National Association of Mutual Casualty Companies, 1958. Data in Appendix F are more recent and may have been determined on a slightly different basis.

Table 3-6 illustrates this useful concept regarding solvent safety. The tolerance concentration in parts per million is given in one column and an evaporation time in another: a low evaporation time signifies a fast-evaporating substance, and low tolerance, a highly toxic material. The *product* of the evaporation time times the tolerance gives us a number that is a better indication of the relative safety than consideration of tolerance alone. The lower the safety number, the more dangerous it is to use the particular solvent. Here we may illustrate the cited example of diacetone alcohol. Because of its low evaporation rate, it can be used with greater safety than chloroform. Again, a far lower concentration of turpentine vapors can be tolerated than acetone. However, due to its high evaporation rate acetone represents a greater danger to the operator using it; note the safety number of acetone, relative to that of turpentine. The hazards of turpentine have been reported in some detail.[31]

Representatives of the Industrial Hygiene Foundation at Mellon Institute who originally pointed out this concept called particular attention to the low safety number and low tolerance of benzene. (Compare its tolerance and safety number with that of a compound of better known toxicity, chloroform.) Although benzene is a very useful solvent in the chemical laboratory, from our knowledge of the homologous series we know that toluene will serve as a ready substitute for many of the things for which benzene can be used. Advisors in matters of industrial safety recommend that toluene or xylene be used in preference to benzene wherever possible; a marked improvement in the safety is obtained.

Several authorities consider that the following classifications present a useful summary of the question of solvent toxicity:

> Low Toxicity
> Paraffinic hydrocarbons
> Ethyl alcohol and higher homologs
> Ethyl, propyl, butyl acetates
>
> Dangerous
> Benzene
> Methanol
> Chlorinated hydrocarbons
>
> Intermediate
> All other familiar solvents

SUMMARY

The subject of solvency and solvent action is too extensive to be covered adequately in a short space. In these brief remarks we have drawn attention to the fact that a solution is the limiting case of dispersing one material in another. The conservator should be aware of the other degrees of dispersion that he often employs: colloidal and coarse dispersion. The chemistry of the colloidal state is an important and extensive subject in itself.

Solvents may simply be classified according to their chemical type. This classification, although fundamental, is not particularly informative regarding evaporation rates or solubility. The solubility characteristics may be measured by a number of empirical methods, such as, the *solubility grade* described here, or various other determinations of *cloud points* and *clear points*. A theoretical basis for solvent action is founded upon the concept of molecular volume and cohesive energy density, expressed in terms of a solubility parameter. The action of solvents is complex; molecular volume and solubility parameter do not explain every case, and new concepts have been developed to take into account dipole forces and hydrogen bonding tendencies.

Tables of evaporation rates are based upon evaporation under specific conditions, such as, evaporation from an open dish or from filter paper. Where such data exist, the information is useful. It is well known that the evaporation rate is not strictly related to the boiling point; yet the boiling range is a customary specification, and, hence, it remains one of the principal properties used in the classification of solvents.

REFERENCES AND NOTES

1. In this connection, it may be mentioned that beeswax does not dissolve to the extent of more than about 8 per cent in turpentine or V. M. and P. naphtha at room temperature. The conservator frequently prepares suspensions of waxes; the amount that remains in solution at room temperature is usually small.
2. L. Scheflan and M. B. Jacobs, *The Handbook of Solvents* (New York: Van Nostrand, 1953).
3. C. D. Bogin in J. J. Mattiello, *Protective and Decorative Coatings I* (New York: Wiley, 1941): Chapter 27, "Lacquer Solvents and Formulation of Solvent Mixture," p. 643.
4. C. M. Hansen, "A Mathematical Description of Film Drying by Solvent Evaporation," *J. Oil & Colour Chemists' Assoc.* 51 (1968): 27-43.
5. W. W. Reynolds, *Physical Chemistry of Petroleum Solvents* (New York: Reinhold, 1963); see Reynolds and co-workers, "The Calculation of the Evaporation Rate of Hydrocarbon Solvents from ASTM Distillation Data," *Official Digest, Federation of Societies for Paint Technol.* 32 (1960): 1146-64; "The Evaporation of Solvents from Thin Films. Theory, Method of Measurement, and Application," *Official Digest, Federation of Societies for Paint Technol.* 33 (1961): 921-39; "The Fundamentals of Solvent Selection," *Official Digest, Federation of Societies for Paint Technol.* 34 (1962): 311-33.
6. H. Burrell, *Interchemical Review* 14, No. 3 (1955): 31; "Solubility Parameters for Film Formers," *Official Digest, Federation of Societies for Paint Technol.* 27 (1955): 726-58; "A Solvent Formulating Chart," *Official Digest, Federation of Societies for Paint Technol.* 29 (1957): 1159-73.
7. P. A. Small, "Some Factors Affecting the Solubility of Polymers," *J. Applied Chem.* 3 (1953): 71-80.
8. J. Hildebrand and R. Scott, *The Solubility of Nonelectrolytes*, 3rd ed. (New York: Reinhold, 1949), pp. 129, 361.
9. R. F. Boyer and R. S. Spencer in H. A. Robinson, "Some Thermodynamic Properties of Slightly Cross-linked Styrene-Divinyl Benzene

Gels," *High Polymer Physics* (Brooklyn, New York: Remsen Press, 1948), p. 476.
10. Taken from a review by R. L. Feller, "Solubility Parameter," *Bulletin of the American Group-IIC* 8, No. 2 (1968): 20-24.
11. J. L. Gardon, "The Influence of Polarity Upon the Solubility Parameter Concept," *J. Paint Technology* 38 (1968): 43-57.
12. J. D. Crowley, G. S. Teague, and J. W. Lowe, "A Three-Dimensional Approach to Solubility," *J. Paint Technology* 38 (1966): 269-80; "A Three-Dimensional Approach to Solubility II," *J. Paint Technology* 39 (1967): 19-27.
13. C. M. Hansen, "The Three-dimensional Solubility Parameter—Key to Paint Component Affinities: I. Solvents, Plasticizers, Polymers, and Resins. II. Dyes, Emulsifiers, Mutual Solubility and Compatibility, and Pigments. III. Independent Calculation of the Parameter Components," *J. Paint Technology* 39 (1967): 104-17; 505-10.
14. J. P. Teas, "Graphic Analysis of Resin Solubilities," *J. Paint Technology* 40 (1968): 19-25.
15. This represents the range indicated in Table 3-2 and more formally shown by Hansen in Figure 3-3.
16. R. Kokes and F. Long, "Diffusion of Organic Vapors into Polyvinyl Acetate," *J. Am. Chem. Soc.* 75 (1953): 6142-46.
17. I. Graham, "The Effect of Solvents on Linoxyn Films," *J. Oil & Colour Chemists' Assoc.* 36 (1953): 500-6.
18. S. Raynolds, "The Dependence of Physical Properties on Chemical Constitution of Alkyl Polymethacrylate Ester Resins" (M.S. thesis, University of Pittsburgh, 1954).
19. R. L. Feller, "Identification and Analysis of Resins and Spirit Varnishes," in *Application of Science in Examination of Works of Art* (Boston: Museum of Fine Arts, 1959), pp. 51-76.
20. S. R. Kiehl, "Evaluation of the Kauri-Butyl Alcohol Solvency Test (for Varnish Thinners)," *Am. Paint and Varnish Manufacturers Assn., Circular 319* (1927): 585; H. A. Gardner and G. G. Sward, *Physical and Chemical Examination of Paints, Varnishes, Lacquers, and Colors*, 12th ed. (1962), p. 480, Kauri-Butanol Test.
21. W. von Fischer, "Hydrocarbon Thinners," *Paint and Varnish Technology* (New York: Reinhold, 1948), p. 231.
22. L. Mandik, "Inter-dependence Between the Kauri-Butanol Value and the Solubility Parameter," *Chem. Prumsyl* 15/40, No. 5 (1965): 282-87; also abstracted in *Review* (Teddington) No. 288 (1966): 477; L. Mandik and J. Stanek, "Selection of Solvents for Coating Systems with the Aid of the Solubility Parameter," *Chem. Prumsyl* 15/40, No. 4 (1965): 223-26.

23. "Freon Fluorinated Hydrocarbon Solvents," *Freon Solvent Bulletin FS-1* (Wilmington, Delaware: E. I. duPont de Nemours and Co., Inc., Freon Products Division, 1957).
24. I. Mellan, *Industrial Solvents* (New York: Reinhold, 1950).
25. G. W. Waters, *Paint Thinners from Petroleum*, 2nd ed. (New York: Shell Oil Company, 1954), p. 36.
26. R. L. Feller, "New Solvent-type Varnishes," in *Recent Advances in Conservation* (London: Butterworths, 1963), pp. 171-75.
27. L. C. F., "How Poisonous are Paint Solvents?" *Industrial Finishing* 35, No. 12 (1959): 81, 94.
28. R. Mallary, "The Air of Art is Poisoned," *Art News* 62 (October 1963): 34-37, 60-61.
29. T. H. Durrans, *Solvents* (London: Chapman and Hall, Ltd., 1950).
30. C. Marsden, *Solvents Manual* (New York: Elsevier, 1954).
31. Anonymous, "Turpentine and Its Hazards," *Paint Manufacture* 30 (1960): 343-44.

Part II

SOLVENT ACTION

Nathan Stolow

4
SOLVENT ACTION: SOME FUNDAMENTAL RESEARCHES INTO THE PICTURE-CLEANING PROBLEM

INTRODUCTION

A varnish coating on a picture surface serves two basic purposes. The prime purpose, well understood and appreciated by the earliest painters, is to improve the appearance of the picture by bringing out more intensely the colors of the pigments. By varying the composition of the varnish and the method of application the artist could also create different degrees of gloss or mattness, thereby contributing further to the aesthetic enjoyment of the work. The second purpose, perhaps more clearly understood in recent years, is the utilization of the protective properties of the varnish. A varnish film, even when thinly applied, does offer some degree of protection to the paint layers against the deteriorating action of oxygen, atmospheric pollutants, and the ever-present dirt and grime of the air. Varnish films also can reduce the photochemical action of light upon pigments and their binding media by partially filtering out the ultra-violet and near-ultra-violet wavelengths. This is certainly the case with yellowed varnish films. Unfortunately, no varnish has yet been found or devised which adequately fulfills the aesthetic and physical requirements and is also absolutely durable. Varnishes, like other organic substances, are subject to the laws of chemical deterioration.

The traditional varnishes used by the artist and the restorer have been composed, essentially, of natural resins, such as, dammar or mastic, dissolved in alcohol (spirits of wine, consisting of 80-90% ethyl alcohol) or distilled turpentine. These varnishes had admirable aesthetic properties because of the way they saturate the colors of a painting and the quality of finish produced. However, they are notorious for their tendency to yellow and become embrittled with age, being chemically altered by oxygen, light, and other environmental factors. It is this yellowing and the degradation of gloss in mastic and dammar varnish films which have prompted restorers to clean pictures and to revarnish them periodically. The damage to these works of art resulting from such frequent cleaning is probably greater than would occur if they were simply allowed to remain with their yellowed and dirt-encrusted varnish intact. The dangers of cleaning with caustic soda mineral acids, sand, and other materials were mentioned by earlier writers, for example, Watin[1] and Déon[2]—mention also being made of these in the 1853 report on the National Gallery, London.[3,4] The early literature on cleaning of pictures is however very sparse, and has been ably summarized by Marijnissen[5] and Ruhemann.[6] A reading of the recipes for cleaning in the earlier writings and our knowledge of the general condition of paintings which have survived the indignities of past cleanings make it not difficult to imagine why a major task of the restorer has often been to repaint and reconstruct. Many errors in cleaning were also to be covered up with tinted varnishes, a practice which reached its peak in the nineteenth century but which still persists today.[3]

The problem of varnish removal has been complicated considerably in those pictures which have been cleaned or partially cleaned several times and where varnishes containing other components have been applied. Thus varnishes containing, besides the traditional mastic or dammar, drying oils, drier salts, and actual pigmented or glaze-like materials (as in restorations and overpainting) make cleaning operations more hazardous and time-consuming. Where the artist has incorporated resins in his painting technique, as is frequently the case, the cleaning agents used for dissolving away the varnish may certainly affect or even remove the paint layers.

While much cleaning in the past, as indeed is still the case, has been for cosmetic purposes, there are times when the varnish must in any case be removed. This is certainly true when works of art must be preserved by lining or transfer processes, conservation techniques designed for consolidating the support.

Whatever the motivation or reason was for picture cleaning in the past, or may be in the present, it is obviously worthwhile to investigate in depth the way in which cleaning agents do their work, and particularly how they may affect the painting itself. Of prime interest are those chemical cleaning agents known as solvents. The possible dangers of solvent action on the original paint film were effectively indicated first in the mid-nineteenth century by such writers as Déon (loc. cit.) and the eminent scientist Faraday.[7] In the published accounts of this period, it was noted that alcohol may cause damage to the paint film. The use of a "restrainer," a second, less active solvent, such as turpentine together with alcohol, probably originated in this period. When the cleaning action using alcohol was too quick, the restorer would flood the surface with the restraining solvent, thereby damping the bite of the alcohol. The accounts also suggest that other liquids, such as linseed oil, nut oil, and coal tar distillates, were sometimes used as the restraining liquid. Owing to their non-evaporating properties, many of these substances would leave a surface film after evaporation of the alcohol. This could cause later problems arising out of migration or absorption into the paint film.

Laurie[8] in 1937, almost a century later, showed by simple experiment that even the alcohol vapors could soften paint. His studies in this regard were in reevaluation of the well-known Pettenkofer process for regeneration of varnish films.[9] In this technique the crazed or dulled varnish could be rendered more or less transparent by exposure to the vapors of absolute alcohol in an enclosed atmosphere. The varnish would still be yellow, but with absorption of the alcohol into the varnish film some physical change would occur that would give greater optical clarity to the painting beneath. This measure is best described as temporary in nature. Ruhemann (loc. cit.) refers to the earlier work of Forni[10] who perhaps anticipated Pettenkofer in certain respects, and who was to lay the groundwork for the re-forming method of Elizabeth Jones.[11] Forni remarked that the greatest advantage of the use of alcohol vapor is not that it regenerates the old varnish, but rather that it makes it easy to remove, "merely with turpentine and without damage to the picture." In 1936, Stout[12] described some interesting trial experiments on the solubility of copal, dammar, mastic, and linseed oil films, which showed that exposures of very short duration to solvents such as acetone, alcohol, and ethylene dichloride cause appreciable film losses. He treated the solubility of varnish films in the light of physical chemistry, demonstrating that swelling is an intermediate stage of solvent action, and that even linseed oil films in this state could suffer from mechanical abrading action. Laurie, in a different line of investigation, found that old films of linseed oil and stand oil could absorb quantities of such viscous materials as bromonaphthalene and paraffin oil.[13] This suggested the ease of penetration of liquids and solvents into such films. Although Laurie did not carry forward this study into the area of solvent action and the resultant mechanical behavior, it is regarded, at least, as a significant effort, and together with the work of Stout marks the beginning of objective research in this field.

The cleaning of pictures, from the empirical methods of restorers, has slowly evolved toward more careful techniques. Unfortunately, the experiences with new solvents or mixtures, and new methods of application, have been recorded inadequately or not at all. It has often been stated that safety in cleaning can lie only in the hands and touch of the restorer and in his trained judgement. Nevertheless, certain tests and safeguards have been devised in the past few decades. Laurie suggested that, instead of using two solvents, alcohol and turpentine (as restrainer), which is subject to certain dangers and uncertainties, a third solvent, known as the diluent, be used. Thus, a typical trio would be alcohol as solvent, xylene as diluent, and turpentine as restrainer. Ruhemann recommended great caution in applying solvents to pictures, noting the many variables involved, and proposed a "safety margin test," in which greater control of solvent action could be achieved.[15] The cleaner has also developed optical controls, such as ultra-violet fluorescence, which can, under certain circumstances, differentiate between varnish and underlying paint. He also has come to employ radiography and microscopy to demarcate areas of repainting or restoration. But it may be said, generally, that there are not many objective criteria or tests which can give the picture cleaner complete confidence and a degree of safety greater than hitherto.

Rees Jones stressed the need for further fundamental research into the entire question of solvent action.[16] This was the period of the Cleaned Pictures Exhibition at the National Gallery, London[17] and of the Weaver Report[18]; much international interest was aroused around the subject of how pictures are, in fact, cleaned. Work on the experimental basis and theoretical interpretation of solvent action on drying oil films, of which linseed oil is the most important, was commenced by the author in 1952 and reported as a doctoral thesis at the University of London.[19] In 1953 Graham carried out some trial experiments on dried oleoresinous varnish films and on dried linseed stand oil.[20] He also showed that solvents enter films by diffusion and cause swelling, and that the degree of swelling at equilibrium depends on the solvent. He suggested that a mixture of turpentine and alcohol, and certain other solvent combinations may cause greater swelling action than the individual solvents acting alone.

The first edition in 1959 of "On Picture Varnishes and Their Solvents" incorporated the author's researches and views on the subject of solvent action up to that date. Subsequently, considerably more work has been carried on by him,[21,22] which necessitated the enlargement and up-dating of this discussion. This new work must also be joined by additional data on the properties of drying oil films obtained by more recent investigations. It is only by such objective studies that further progress can be made in this difficult subject of solvent action. The following sections essentially cover the same topics as before, including original data and the more recent results.

SOME REMARKS ON THE DRYING OF LINSEED OIL FILMS

It would be worthwhile to review some of the properties of linseed oil in that it is the most common constituent in oil paintings. Linseed oil before

FUNDAMENTAL RESEARCHES INTO PICTURE CLEANING

drying consists largely of triglycerides of five fatty acids, namely, palmitic, stearic, oleic, linoleic, and linolenic acids. Typical triglyceride molecules in unreacted linseed oil may be of these types:

(a)	(b)	(c)
⎡Stearic ⎢Palmitic ⎣Oleic	⎡Palmitic ⎢Oleic ⎣Linoleic	⎡Linoleic ⎢Linolenic ⎣Linolenic

where the fatty acids are joined to the glycerol backbone. A more correct formula representation of (a) is:

$$\begin{array}{l} \text{H}-\underset{\underset{\text{H}}{|}}{\overset{\overset{\text{H}}{|}}{\text{C}}}-\text{O}-\overset{\overset{\text{O}}{\|}}{\text{C}}-(\text{CH}_2)_{16}-\text{CH}_3 \\ \text{H}-\underset{|}{\text{C}}-\text{O}-\overset{\overset{\text{O}}{\|}}{\text{C}}-(\text{CH}_2)_{14}-\text{CH}_3 \\ \text{H}-\underset{\underset{\text{H}}{|}}{\text{C}}-\text{O}-\overset{\overset{\text{O}}{\|}}{\text{C}}-(\text{CH}_2)_7-\text{CH}=\text{CH}-(\text{CH}_2)_7-\text{CH}_3 \end{array}$$

The distribution of the fatty acids as ester components in the linseed oil triglycerides generally conforms to a modified random distribution.[23,24,25,26] Table 4-1 relates linseed oil composition to triglyceride species based on such studies.

TABLE 4-1. TRIGLYCERIDE COMPOSITION OF LINSEED OIL (MOLE %)

DUTTON & CANNON[23]		GUNSTONE & PADLEY[25]		VERESCHAGIN & NOVITSKAYA[26]	
*333	18%	333	22%	333	16.2%
332	12	332	15	332	11.7
331	20	331	18	331	14.9
322	4	330	10	330	10.5
Unspecified	46	322	4	322	3.5
		321	8	321	7.9
		320	5	320	5.8
		310	6	310	10.3
		300	1	300	0.0
		Other	11	Other	~20.0

*Here the numbers 3,2,1,0 refer respectively to the linolenic, linoleic, oleic, and saturated acids (either palmitic or stearic) as functional parts of the triglycerides in question.

The ability of linseed oil and other drying oils to dry to hard films is attributed to the presence of unsaturated acid components in the triglyceride molecules, e.g., oleic, linoleic, and linolenic acids with respectively one, two and three double bonds ($-\overset{|}{C}=\overset{|}{C}-$). The drying capacity of linseed oil depends on the relative amount of the more unsaturated acid components, e.g., linolenic, and the total amount of all unsaturated species. A useful index for determining "drying" power of an oil is the iodine value (I.V.),[27] which is a measure of total unsaturation of the triglycerides originally present. The differentiation between drying and non-drying oils based on iodine values is roughly as follows:

Drying oils, I. V. 200–160–e.g., linseed, parilla, china wood oil	Rich in linolenic, and moderately rich in linoleic acids.
Semi-drying oils, I. V. 150–120–e.g., poppy seed, walnut, safflower, tobacco seed oils.	Low in linolenic, rich in linoleic acids.
Non-drying oils, I. V. less than 100–castor, olive, cottonseed oils.	Rich in linoleic and oleic acids, as well as saturated acids.

The process or mechanism of film formation has been the subject of much research and speculation over the years. The more recent views postulate that drying oils may form films in various ways depending on temperature, exposure to oxygen and to light.[28] Above about 40°C. thermal polymerization reactions predominate. At room temperatures the drying oils undergo a process of autoxidation when exposed to air. The initial degree of unsaturation of the oil has a great bearing on the drying process. The formation of hydroperoxide links at the active sites in linolenic, linoleic (and to a lesser extent oleic) acid portion of the triglyceride molecules has been suggested as one of the early stages of polymerization. These active sites are at the carbon atoms between the double bonds. The thickness of the film also has a bearing on the manner of drying, for the lower concentration of oxygen present in thick films promotes different polymerization reactions than in thin films. Owing to the presence of such acid components as palmitic and stearic in a portion of the triglycerides of the starting oil, which are very unreactive because of their saturation (i.e., lack of double bonds), they may remain unchanged after the film is dry. Indeed there is evidence to show that unreacted triglycerides do remain in the film and are extractable by solvent. This will be dealt with shortly. The drying process results, not only in the formation of a mesh of triglycerides bonded together with unreacted material, but also in breakdown products. These breakdown products arise out of the fission, or scission, of molecules, at possibly different reactive sites than before, to yield a variety of small molecules. This is shown schematically in Figure 4-1. Some of the more

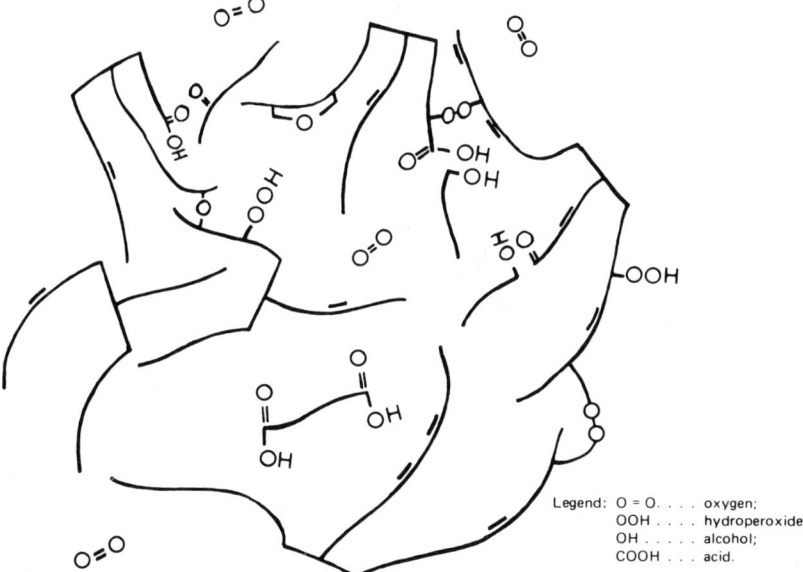

Figure 4-1 Breakdown of Hydroperoxide and Other Polymeric Linkages in an Aged Dried Linseed Oil Film. A variety of smaller molecules (e.g., dicarboxylic acids) may result from scission reactions.

significant of these are dicarboxylic acids of six, seven, eight, or nine carbon atoms, e.g.:

$$HO-\underset{}{\overset{O}{\overset{\|}{C}}}-\underset{H}{\overset{H}{\overset{|}{C}}}-\underset{H}{\overset{H}{\overset{|}{C}}}-\underset{H}{\overset{H}{\overset{|}{C}}}-\underset{H}{\overset{H}{\overset{|}{C}}}-\underset{}{\overset{O}{\overset{\|}{C}}}-OH \quad \text{Adipic acid}$$

$$HO-\underset{}{\overset{O}{\overset{\|}{C}}}-\underset{H}{\overset{H}{\overset{|}{C}}}-\underset{H}{\overset{H}{\overset{|}{C}}}-\underset{H}{\overset{H}{\overset{|}{C}}}-\underset{H}{\overset{H}{\overset{|}{C}}}-\underset{H}{\overset{H}{\overset{|}{C}}}-\underset{}{\overset{O}{\overset{\|}{C}}}-OH \quad \text{Pimelic acid}$$

$$HO-\underset{}{\overset{O}{\overset{\|}{C}}}-\underset{H}{\overset{H}{\overset{|}{C}}}-\underset{H}{\overset{H}{\overset{|}{C}}}-\underset{H}{\overset{H}{\overset{|}{C}}}-\underset{H}{\overset{H}{\overset{|}{C}}}-\underset{H}{\overset{H}{\overset{|}{C}}}-\underset{H}{\overset{H}{\overset{|}{C}}}-\underset{}{\overset{O}{\overset{\|}{C}}}-OH \quad \text{Suberic acid}$$

$$HO-\underset{}{\overset{O}{\overset{\|}{C}}}-\underset{H}{\overset{H}{\overset{|}{C}}}-\underset{H}{\overset{H}{\overset{|}{C}}}-\underset{H}{\overset{H}{\overset{|}{C}}}-\underset{H}{\overset{H}{\overset{|}{C}}}-\underset{H}{\overset{H}{\overset{|}{C}}}-\underset{H}{\overset{H}{\overset{|}{C}}}-\underset{H}{\overset{H}{\overset{|}{C}}}-\underset{}{\overset{O}{\overset{\|}{C}}}-OH \quad \text{Azelaic acid}$$

54   SOLVENT ACTION

Films containing pigment have somewhat different drying characteristics. Some pigments containing lead or cobalt are known to accelerate the drying process, while others may retard the action. The presence of pigment particles may also slow down diffusion processes and in some cases reduce the action of light on the linseed oil portion of the film. These factors will be considered further on; all have a bearing on understanding the process of solvent action on films containing dried linseed oil.

LEACHING AND SWELLING ACTION OF SOLVENTS

When a dried film is exposed to solvent for the first time, a marked action is observed. The solvent rapidly penetrates into the film, swelling the outer portions first and then, given sufficient time, swelling the entire film. If the film is supported on a rigid and impermeable surface, the solvent penetration will cause the film to expand in thickness only, unless the film-support bond is ruptured. If the film is unsupported the swelling will be evident in all three dimensions. Almost as soon as solvent penetrates the film, soluble material locked in the film structure (discussed in the previous section) starts to diffuse out of the film. After a period of time no more soluble material will be extracted, although the film will still be swollen. It is interesting to observe

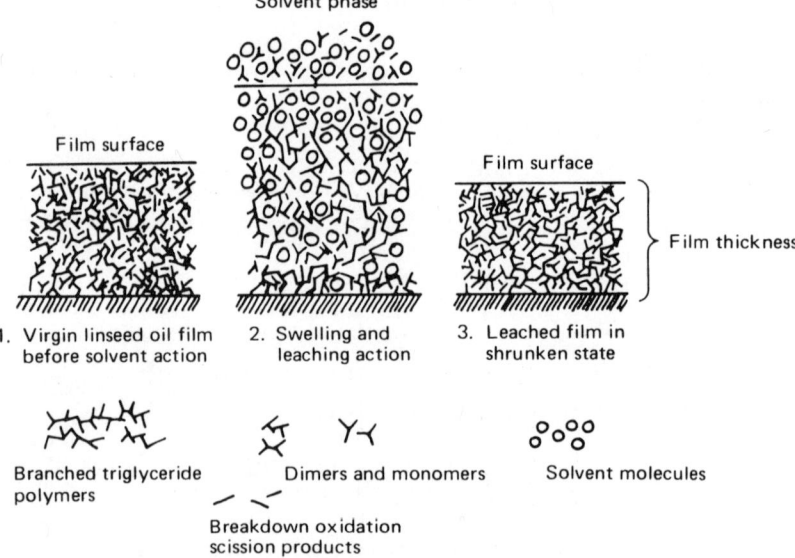

Figure 4-2 Schematic Representation of Swelling, Leaching, and De-swelling Processes.

that during the stage of the extraction of soluble material, referred to as *leaching*, the swelling volume of the film decreases progressively to a static value as long as solvent remains in contact with the film. When the solvent is allowed to evaporate away—by removing the swollen film—the film commences to deswell and ends with a final volume which is substantially less than the previous one. In terms of physical characteristics, the film is then denser, more brittle, and may possibly be deformed out of plane. The schematic representation of the swelling, leaching, and deswelling processes is shown in Figure 4-2. Described in Figure 4-3 is the way in which an

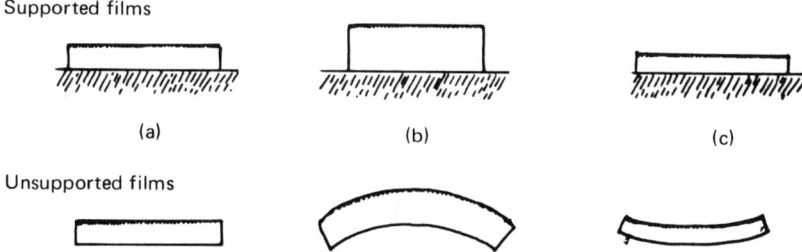

Figure 4-3 Manner of Swelling and De-swelling of Supported and Unsupported Films. (a)–original film (the top side of film indicated by heavy line); (b)–swelling action causing convex distortion of unsupported film; (c)–deswelling results in concave distortion of unsupported film.

unsupported film may become distorted upon solvent contact, particularly if the film is thick and has a nonuniform structure. The actual experimental results are given in the graphs of Figure 4-4. Here the swelling measurement has been followed by dimensional measurement as a function of time of solvent contact. The supported films have been measured on a special swelling measurement apparatus based on the principle of the parallel plate viscometer, while the unsupported films have been measured by means of a traveling microscope.[29]

INITIAL SWELLING/LEACHING PROCESS

Consider the action of a solvent (acetone) upon a supported film of white lead/stand oil paint, as shown in Figure 4-4(a). This film is supported on a glass plate, aged for thirty-five weeks at 32°C., and of thickness 105μ (microns). As soon as acetone is brought into contact with the film, swelling occurs, as shown by the increased thickness measured by the swelling measurement apparatus. In just a few minutes the swelling curve levels off to a

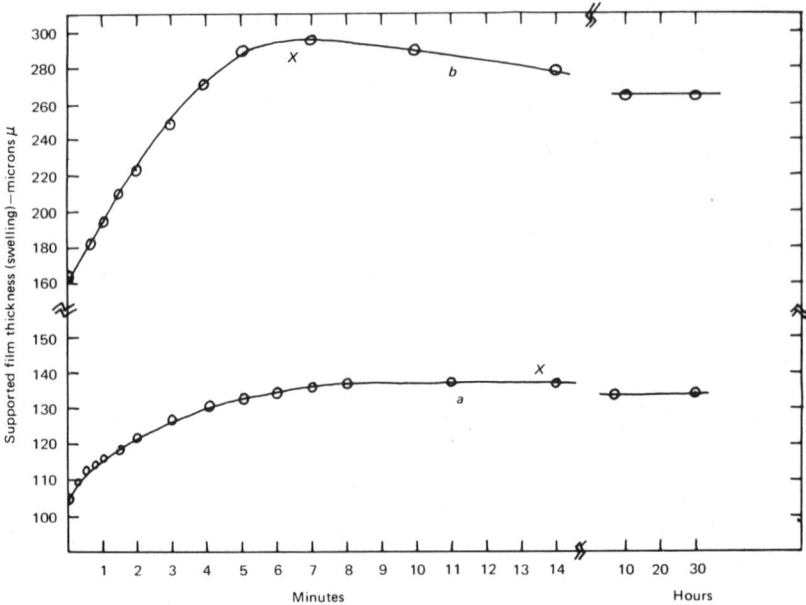

Figure 4-4 The Initial Action of Solvent upon Supported Films. (a) White lead/stand oil glass supported film aged for thirty-five weeks at 32°C., 105μ thick, brought into contact with acetone at 22°C.
(b) Stand oil glass supported film aged for twenty-four weeks at 32°C., 162μ thick, brought into contact with acetone at 22°C.

maximum (at X), then slowly drops to an equilibrium swelling thickness of 134μ. In the interval up to X the film imbibes solvent very rapidly, becoming swollen, and at the same time soluble matter leaches out, the more so as the film structure expands. Beyond X the swelling action is overtaken by leaching, and the swollen film shrinks steadily until the leaching process is completed. It is important to note that, although the leaching process is completed, the film remains swollen, though at a lower level. As is shown in the next section, the film on evaporation of the solvent shrinks to smaller dimensions than before. In unpigmented films (e.g., those in Figure 4-4[b]) the swelling/leaching curves peak to a maximum more sharply. In unsupported or free films similar behavior is to be noted. Here both sides of the film are exposed to solvent and the swelling/leaching action occurs more quickly than in the supported films. Thus the maxima are reached more quickly, and the curves fall off more quickly during the final leaching stage (Figure 4-5, [a] and [b]). The pigmented films in Figure 4-5(a) have been aged for seven years under normal room temperature conditions before solvent treatment. The

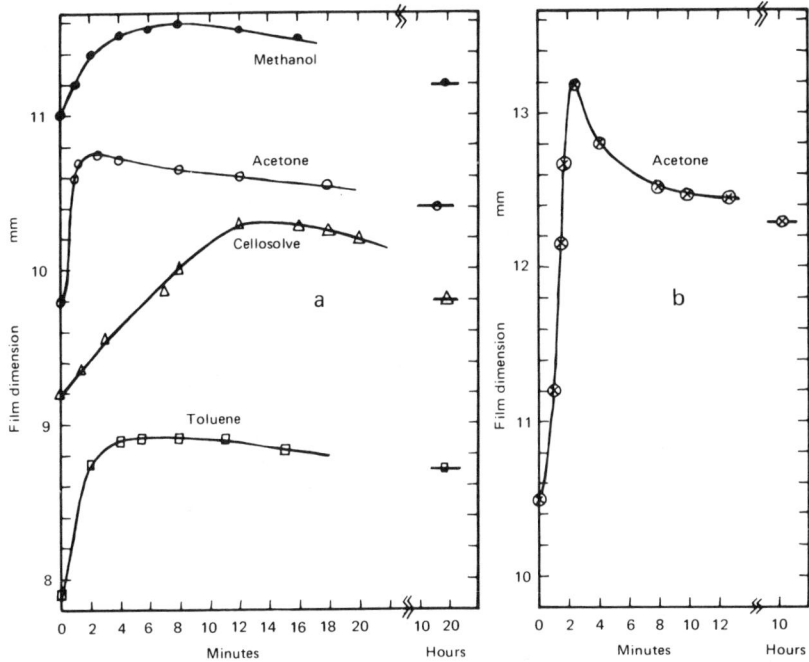

Figure 4-5 The Initial Action of Solvent upon Unsupported Films. (a) White lead/stand oil films aged for seven years 20-30°C., 130-140µ thick, brought into contact with solvents indicated at 22°C. (b) Stand oil film aged for seven years 20-30°C., 125µ thick, brought into contact with acetone at 22°C. In both (a) and (b) swelling and leaching action followed by measuring dimensional changes of the films.

behavior of the corresponding unpigmented films of the same age (Figure 4-5[b]) is essentially the same. The thickness of the film and the degree of pigmentation (that is, the ratio of pigment to medium) would undoubtedly influence the shape of the swelling/leaching curve. The effect of changing the solvent is to shift the maximum of the swelling/leaching curve to lesser or greater times, depending on whether the solvent diffuses quickly or slowly into the film structure. Thus, for the white lead/stand oil seven-year-old films studied (Figure 4-5[a]) the maxima for acetone and cellosolve occur at 2.5 min. and 14 min., respectively. Acetone is of course a much faster diffusing solvent than cellosolve. In the unsupported or free films the swollen film eventually reaches a lower level, and after evaporation shrinkage occurs, but this time in all dimensions. In this respect the free films differ from the supported films, which swell and shrink in one plane only. The swelling/leaching phenomenon has been observed consistently to be the same as described above, regardless of film type, age, thickness, or solvent used. Upon first contact with solvent, swelling and leaching occur together, and after a certain time the leaching process predominates.

Some limited experiments were carried out with varnish-coated paint films to see whether the presence of a surface film would modify the usual swelling/leaching behavior. The action of acetone on dammar-coated white lead/stand oil films aged for seven years are given in Figure 4-6. For comparison

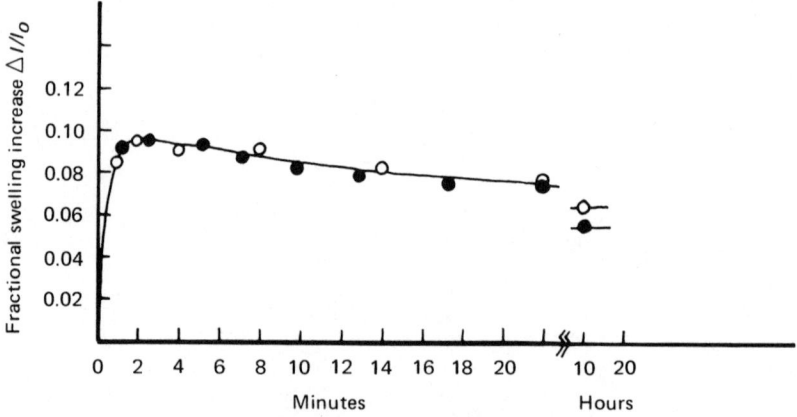

Figure 4-6 Comparison of Action of Acetone on Dammar Coated (–●–) and Uncoated (–○–) Films. The films are white lead/stand oil unsupported, aged for seven years 20-30°C., 132μ thick (with dammar coating, 142-μ thick).

the solvent action on the uncoated film is included. It is seen that the dammar film (10μ thick) is penetrated so easily and quickly by the acetone that the course of swelling and leaching is virtually unaffected. In a matter of a minute or less the surface coating would be dissolved away. Earlier studies of solvent action on supported dammar and mastic films using the aforementioned swelling measurement apparatus showed that a film coating of the order of 10-20μ thick would be penetrated by acetone or methyl alcohol in seconds, and this would be followed very quickly by dissolution. The application of thicker and more solvent-resistant coatings to a paint film would merely slow down the swelling/leaching process, but would not eliminate it entirely.

Consider now what happens when the solvent-treated films are allowed to evaporate. They will now occupy less volume, will be reduced in weight, and will be of increased density. This has been borne out experimentally and supports the schema described in Figure 4-2.

TABLE 4-2. FILM SHRINKAGE AND INCREASE IN DENSITY RESULTING FROM SWELLING/LEACHING ACTION.

FILM TYPE	FILM AGE & TEMPERATURE	SOLVENT	SHRINKAGE (%)	DENSITY BEFORE LEACHING (gm./cm.^{3})	DENSITY AFTER LEACHING (gm./cm.^{3})	SOLUBILITY (%)
Supported Films:						
White lead/open pot stand oil; 81% pigment, 19% oil content	18 weeks 32°C.	Methanol	14*	3.53	3.82	4.5
White lead/open pot stand oil; 79% pigment, 21% oil content	35 weeks 32°C.	Acetone	7	3.30	3.75	3.2
Linseed oil containing 0.5% Pb as drier	27 weeks 32°C.	Acetone	30	1.145	1.170	24.0
Unsupported Films:						
White lead/open pot stand; oil 79% pigment, 21% oil content	350 weeks 20-30°C.	Acetone	9**	3.40	3.80	4.0
White lead/stand oil; 79% pigment, 21% oil content	1250 weeks 20-30°C	Isopropanol	16	—	—	5.7
White lead/stand oil; 79% pigment, 21% oil content	1250 weeks 20-30°C.	Ethanol	17	—	—	5.2
Vacuum stand oil	6 weeks 65°C.	Acetone	34	—	—	28.2

*In supported films the *shrinkage* is calculated on the basis of decrease of film thickness, after leaching, from the unleached thickness.
**In unsupported films the shrinkage is calculated on a *volume basis*. The dimensional changes are noted and from these the volume decrease is calculated.

The *solubility* is based on the weight of the starting film. The solubility figures for the pigmented films are of a lower level as a result. If the pigment/medium concentration was taken into account, the solubility figures would be brought into line with those for the unpigmented films.

Some typical data for film shrinkage and density changes are given in Table 4-2. It will be seen that, in all cases referred to, the effect of the swelling/leaching treatment of films, whether pigmented or not, is to produce a shrunken, denser structure. As a result of losing soluble material the films are less flexible, i.e., more brittle. It appears that the soluble components function as a kind of film plasticizer. Some calculations based on the experimentally observed parameters of density and solubility show that the leached components have densities of the order of the original oil of the film. This is in accordance with the idea that the leached materials are of lower molecular weight species. It is interesting to note that the pigmented white lead films are quite strongly structured, suffering no pigment loss on solvent treatment, notwithstanding the strains imposed by swelling and shrinking. However, some other film types studied incorporating iron oxide were found to lose particles of pigment upon solvent action. Obviously the factor of pigment/medium interaction enters here, some pigments being strongly coordinated with the film medium and others less so. The unsupported films, again, were observed to shrink in all dimensions, more so in thickness than in length and width. This suggests that there is some preferred structural orientation in the dried film, and that it is somewhat easier for the polymers of the film to stretch vertically than horizontally.

REPEATED SWELLING ON LEACHED FILMS

It is interesting to know how a previously leached film behaves when it is exposed to the same solvent a second time. If the leaching process has been practically complete in the first stage, then the second contact with solvent will be essentially swelling action only. This is illustrated in Figure 4-7 for a few typical types of film. In every case the second swelling shows no maximum peak (cf. Figures 4-4 and 4-5), but rather a smoothly rising curve reaching a maximum level indicating that an equilibrium value of swelling has been reached. It is usual to express the swelling amount as a fraction of the original film volume. Thus the equilibrium degree of swelling is generally expressed as: $\Delta V/V_o$, where ΔV is the maximum change of volume of the film upon swelling and V_o is the starting volume of the film (for example, the initial volume of the leached film). Where the films are supported, the equilibrium degree of swelling may be expressed in equivalent form by $\Delta h/h_o$, where Δh is the maximum change in film thickness and h_o is the starting thickness. For free unsupported films, the values of $\Delta V/V_o$ are calculated from the observed dimensional changes.

The degree of swelling for the unleached films is always observed to be greater than for successive swellings in the same solvent. However, once the film is leached, repeated swellings remain essentially unchanged, provided no

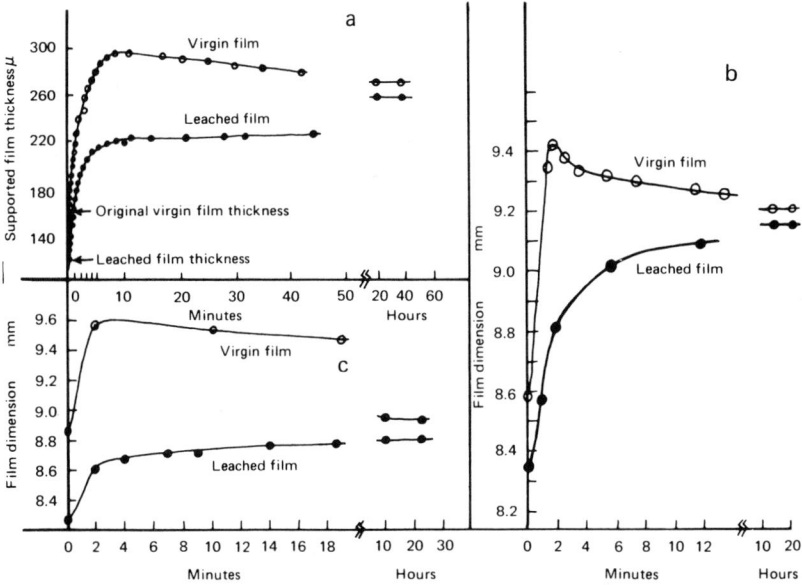

Figure 4-7 Swelling of Virgin and Leached Films. (a) Vacuum-bodied stand oil glass supported film, aged for twenty-four weeks at 32°C., virgin film thickness 162μ, leached film thickness 118μ, solvent: acetone 22°C. (b) White lead/stand oil unsupported, aged for seven years 20-30°C., virgin dimension of selected side: 8.58 mm.; after leaching: 8.35 mm., solvent acetone 22°C. (c) White lead/stand oil unsupported, aged for twenty-five years 20-30°C., virgin dimension of selected side: 8.83 mm.; after leaching: 8.26 mm., solvent acetone 22°C.

further leaching occurs and that the film structure is not modified or altered by other factors, such as further polymerization, or deterioration with increased aging. In Table 4-3 are presented data for some film types illustrating this experimental observation.

Referring again to Figure 4-7, it is seen that once a film is leached the equilibrium degree of swelling (second swelling) is reached quite quickly compared with the virgin films. The convenient indication of "time for half-completion" of the swelling process may be used to estimate the speed of swelling. Thus in the twenty-five-year-old white lead/stand oil films, the half time for acetone is 1.4 min., and for the vacuum bodied stand oil films the half time is about 1.5 min. As will be discussed later in the section on diffusion of solvents into films, the speed or rate of swelling action depends essentially on film thickness, solvent, temperature, presence or absence of

TABLE 4-3. REPEATED SWELLING ACTION ON FILMS—VIRGIN AND LEACHED FILMS (UNSUPPORTED)

FILM TYPE AND AGING	SOLVENT	DEGREE OF SWELLING, VIRGIN FILM, $\Delta V/V_0$	SECOND SWELLING $\Delta V/V_0$
(i) Vacuum stand oil with 0.5% Pb as drier, thickness 140μ. Aged 7 years 20–30°C.	Acetone Methanol Ethanol Methyl Cellosolve	2.54 0.56 1.08 2.05	1.32 0.46 1.05 1.70
(ii) White lead/open pot stand oil; 81% pigment, 19% oil content, thickness 130μ. Aged 7 years 20–30°C.	Acetone	0.32	0.31
(iii) White lead/stand oil; 79% pigment, 21% oil content, thickness 50μ. Aged 25 years 20–30°C.	Acetone	0.25	0.22

pigment, and on whether the film is supported or allowed complete access to solvent on all sides.

LEACHING ACTION AND TIME OF SOLVENT CONTACT

Studies were made of the actual losses of soluble material from films at different times of solvent contact. As mentioned earlier, these losses are indicated in the shape of the swelling/leaching curves, those which show peaks in the plot of swelling versus time. The solubility experiments were carried out on sections of the same type of film exposed to solvent for different time intervals, removing and evaporating off solvent in a high vacuum to constant weight. The amount of leached material would then be obtained from the initial and final film weights. The results for two film types are shown in Figure 4-8. The graphs are marked X to correspond with the maximum peak in the swelling/leaching-versus-time curve. It is significant that the loss of soluble material occurs very rapidly, within seconds of solvent contact. Thus, in both the pigmented and unpigmented films, 50% of the soluble material is removed by acetone in 100 sec. It may be expected that if slower solvents are used, the effect of pigmentation would be to slow down the rate of loss of soluble material. The terminal solubility of 4.0% for the pigmented films is based on total film weight. If corrected for pigment content, this solubility would in fact be $\frac{4.0}{0.207}$, or 19.3%, which is of magnitude similar to that for the unpigmented films.

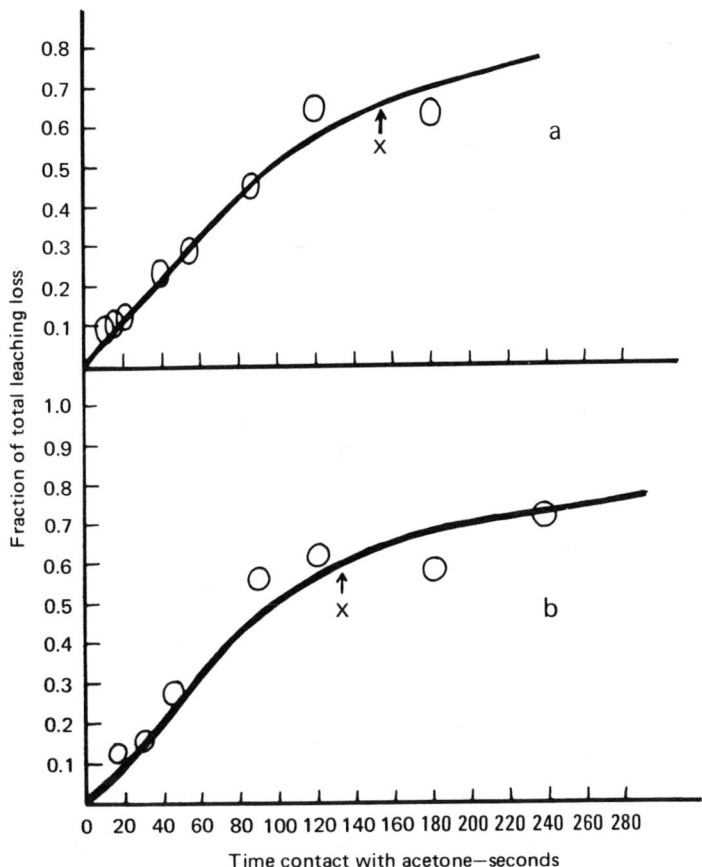

Figure 4-8 The Leaching of Films with Time of Solvent Contact at 22°C. (a) White lead/stand oil, unsupported, aged for seven years 20-30°C., 134μ thickness before leaching, solvent acetone 22°C., maximum solubility 4.0% (based on initial weight of film). (b) Vacuum-bodied stand oil, unsupported, aged for 100 days at 65°C., 100μ thickness before leaching, solvent acetone 22°C., maximum solubility 28.2% (based on initial weight of film).

AMOUNT OF LEACHING: SOLVENT AND FILM TYPE

Here the effect of changing the solvent and the film type upon the degree of extraction of soluble material is considered. Experiments designed to study these variables are summarized in Table 4-4. Comparisons within a film type are for the same age, to avoid introducing another complicating variable. In

TABLE 4-4. LEACHING: SOLVENT AND FILM TYPE VARIATION
(FREE FILMS)

FILM TYPE, MANNER OF AGING	SOLVENT	"FILM" LEACHING (%)	"MEDIUM" LEACHING* (%)	REMARKS
Pigmented Films:				
(i) White lead/stand oil; 79% pigment, 21% oil content, 50μ thick. Aged in diffuse light 25 years 20–30°C.	Acetone Ethanol Isopropanol	5.6 ± 0.8 5.2 ± 0.4 5.7 ± 0.4	24 ± 1.5 25 ± 1.5 27 ± 1.5	Unleached film moderately flexible; after leaching very brittle.
(ii) White lead/stand oil; 81% pigment, 19% oil content, 130μ thick. Aged in diffuse light 7 years 20–30°C.	Acetone Methanol Cellosolve Toluene	4.0 ± 0.3 4.0 ± 0.5 5.0 ± 0.5 3.0 ± 0.5	21 ± 1.5 21 ± 1.5 26 ± 1.5 16 ± 1.5	Unleached film moderately flexible; after leaching not quite so brittle as in (i).
(iii) White lead/alkali-refined linseed oil; 82.7% pigment, 17.3% oil content, 90μ thick. Aged in diffuse light 3.3 years 22°C.	Acetone	4.9 ± 0.5	28 ± 1.5	Unleached film brittle; after leaching very brittle.
(iv) White lead/alkali-refined linseed oil; 84.7% pigment, 15.3% oil content, 90μ thick. Aged under strong ultra-violet light 3.3 years 22°C.	Acetone	3.7 ± 0.5	24 ± 1.5	Unleached film very brittle and hard; after leaching extremely brittle.
(v) Titanium dioxide (rutile)/alkali-refined linseed oil; 72.5% pigment, 27.5% oil content, 90μ thick. Aged in diffuse light 3.3 years 22°C.	Acetone	17 ± 1	62 ± 3	Unleached film flexible; after leaching more rigid.
(vi) Titanium dioxide (rutile)/alkali-refined linseed oil; 75% pigment, 25% oil content, 90μ thick. Aged under strong ultra-violet light 3.3 years 22°C.	Acetone	21 ± 1	84 ± 4	Unleached film disintegrated rapidly upon contact with solvent, liberating pigment particles.

FUNDAMENTAL RESEARCHES INTO PICTURE CLEANING

TABLE 4-4.—Continued

FILM TYPE, MANNER OF AGING	SOLVENT	"FILM" LEACHING (%)	"MEDIUM" LEACHING* (%)	REMARKS
(vii) Iron oxide (Fe_2O_3)/ alkali-refined linseed oil; 84% pigment, 16% oil content, 90μ thick. Aged under strong ultra-violet light 3.3 years 22°C.	Acetone	14 ± 1	87 ± 4	Same kind of film disruption as in (vi).
(viii) Barium sulfate/ alkali-refined linseed oil; 72.3% pigment, 27.7% oil content, 90μ thick. Aged in diffuse light 330 days 22°C.	Acetone Methanol Isopropanol Methylene-chloride Chloroform n-Hexane	14.5 ± 0.5 15.4 ± 0.5 13.8 ± 0.5 12.6 ± 0.5 15.3 ± 0.5 4.7 ± 0.5	52.5 ± 1.5 56 ± 1.5 50 ± 1.5 46 ± 1.5 55 ± 1.5 17 ± 1.5	Unleached films flexible; after leaching less flexible but not brittle.
(ix) White lead/linseed oil; 82.4% pigment, 17.6% oil content, 200μ thick. Aged in diffuse light for approximately 48 years 20–30°C. Artist's palette.	Acetone	3.5 ± 0.4	20.5 ± 1.5	Unleached film brittle and hard; same for leached film.
Unpigmented Films:				
(i) Alkali-refined linseed oil, 80μ thick. Aged under strong ultra-violet light 3.3 years 22°C.	Acetone	27 ± 0.5	27 ± 0.5	Film slightly flexible before leaching, more brittle afterwards.
(ii) Open pot stand oil, 150μ thick. Aged 187 hrs. at 65°C, then 2 years at 22°C., in diffuse light.	Acetone Ethanol Isopropanol Cellosolve Carbon tetrachloride	21 ± 1 26 ± 1 23 ± 1 26 ± 1 21 ± 1	21 ± 1 26 ± 1 23 ± 1 26 ± 1 21 ± 1	Film flexible before leaching (slightly sticky), less flexible and less sticky after leaching.

*The "Medium" Leaching (%) is obtained by dividing the film leaching percentage by the fraction of film occupied by the oil medium. In this way comparison may be made with the leaching for unpigmented films.

every case sufficient time has been allowed for the leaching action to be complete so that comparisons may be made.

The results show that the percentage of extracted material, once corrected for pigment content ("Medium" Leaching Per Cent column), are more or less of the same order for white lead films, unpigmented stand oil, and the alkali-refined linseed oil (aged under strong ultra-violet light). The other pigmented films, titanium dioxide (rutile), iron oxide (Fe_2O_3), and barium sulfate, which are made with alkali-refined linseed oil, exhibit very high solubilities of the order of 50-87% in acetone. These films are obviously not as polymerized or cross-linked as those based on white lead, or those aged under strong ultra-violet light. Obviously the degree of solubility depends on film and aging characteristics. Nevertheless in every case solubility behavior was noted.

A limited number of experiments on samples from paintings and other sources likewise demonstrate the leaching phenomenon. Thus in Table 4-4 (ix) the white lead paint fragments from an artist's palette of circa 1920 are soluble to the extent of 3.5%. Changing the solvent, with the exception of n-hexane, does not alter the leaching percentages appreciably. The low value for n-hexane may be explained by the low swelling power this solvent has for linseed oil films. Obviously the degree of leaching will be influenced by the effectiveness of the solvent in swelling the film. From the tabulated results and from general observations it may be said that paint films of the linseed oil type, irrespective of age, may be leached. In a recent study[30] on identification of media of old paintings, Masschelein-Kleiner et al. have noted that fragments of oil paint from the fifteenth century are sufficiently soluble in chloroform to permit analysis. Reference here is presumably to the leaching action in chloroform. Some pigment-medium combinations will be more resistant to leaching by solvents than others. After leaching action the resulting films are less flexible than before, and in some cases are extremely brittle. This results from removal of components which somehow plasticize the original film. Of the various films made in the laboratory those aged under ultra-violet light were the most brittle to start with and had already lost fair amounts of organic material. Thus the white lead/alkali-refined films, (iii) and (iv), show that in ultra-violet aging for 3.3 years, compared with normal aging, the oil medium content decreases from 17.3 to 15.3% in comparison with the value of 21% in the originally formulated paint. The figures for leaching in the last column of Table 4-4 should be corrected for such compositional changes with aging, but it is to be expected that they would be altered by only a few per cent at most. A precise evaluation of leaching losses should be made in terms of the starting composition when the film was first laid down.

AMOUNT OF LEACHING AS THE FILM AGES

The leaching of three selected types of film has been studied for extensive aging periods, as is illustrated in Figure 4-9. It is seen here that the film solubility falls off rapidly in the early stages of formation (up to approximately twenty weeks), then levels off and appears to rise slowly again. This may readily be explained in terms of the formation and deterioration of

FUNDAMENTAL RESEARCHES INTO PICTURE CLEANING 67

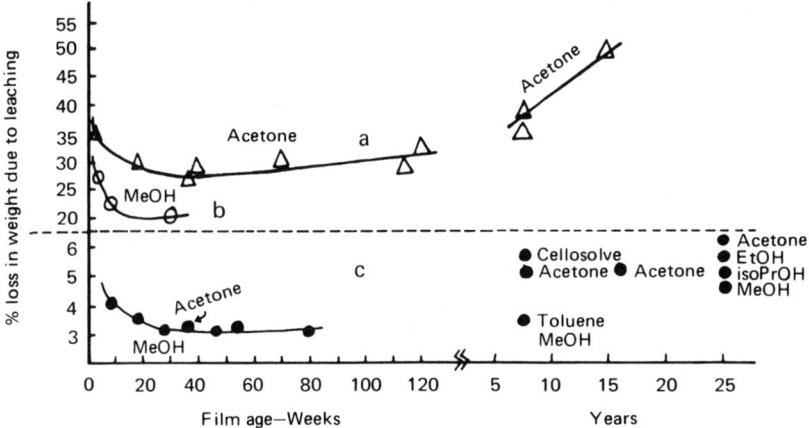

Figure 4-9 The Effect of Film Age on Amount of Leaching. (a) Vacuum-bodied stand oil aged 20-30°C. (b) Vacuum-bodied stand oil aged 20-30°C. (c) White lead/stand oil aged 20-30°C. In each case film in contact with respective solvent for more than 40 hours or until maximum leaching action reached.

linseed type films. In the first several weeks the films become polymerized, and soluble materials (mainly unreacted triglycerides) are removed by solvent action quite readily. During the middle period of film formation polymerization progresses further, but then oxidation processes take their toll on the film polymers. The soluble material in films of greater age would then consist of the triglycerides having low capacity (if any) to polymerize, and the oxidation products. The individual plotted points of Figure 4-9 are for virgin portions of the film in question. Again, as was shown in the preceding section, the leaching figures could be rendered more precise by referring to a fixed original film composition.

The white lead films are observed to dry to great hardness after extensive aging, more so than the other pigmented films, but this does not appear to affect the leaching behavior. The unpigmented vacuum stand oil films do not really harden with extended aging, but on the contrary appear to "de-polymerize," developing a certain stickiness or fluidity. This is even more marked with films formed from ordinary linseed oil. It has been noted that solubilities of the order of 50-80% are obtained for such films after three years or so of ambient aging. This level of high solubility with pigment separation has already been noted for the iron oxide, titanium dioxide, and barium sulfate/alkali-refined linseed oil films of Table 4-4.

In separate experiments previously leached white lead/stand oil films were stored for several years and then brought into contact with solvent again. Further leaching action was recorded. Undoubtedly this is due to continued film deterioration. Some of the results are entered in Table 4-5.

TABLE 4-5. CONTINUED FILM LEACHING AFTER FURTHER AGING

FILM TYPE, MANNER OF AGING	SOLVENT	FILM LEACHING (%)	FURTHER AGING LEACHED FILM (years)	FURTHER FILM LEACHING (%)
(i) White lead/stand oil; 81% pigment, 19% oil content, 200μ thick, Aged 1 year 33°C.	Acetone	4.0 ± 0.5	14	1.5 ± 0.3
(ii) White lead/stand oil; 81% pigment, 19% oil content, 130μ thick, Aged 7 years 20–30°C.	Acetone	5.3 ± 0.5	8	0.7 ± 0.2

Thus after a period of time, measured in years, there is a further build-up of solvent-extractable material in the film structure as a result of continuing oxidative breakdown processes. In (ii) of Table 4-5 a close check was kept on film weights at all times and it was observed that the leached films lying dormant for eight years had gained 1.5% in weight. This can best be explained on the basis of oxygen gain by the organic part of the paint film. Also, in a parallel experiment the leached films of (ii) had been treated with methylene chloride instead of acetone, yielding a further leaching value of 0.8%. This is effectively the same as that for acetone. Analysis of these leachings showed that the dissolved components are practically the same.

THE LEACHED COMPONENTS

Up to now the various parameters affecting film solubility have been dealt with in their turn. It is now appropriate to investigate the nature of the leached components. The identification of these components has been aided by modern analytical techniques employing the infra-red spectrophotometer, the gas-liquid chromatograph, and thin-layer chromatography. Portions of this work have been described in some detail by the author, and the instrumental techniques need not be discussed in detail here. [22] Earlier, it was said that the leachings are expected to contain unpolymerized triglycerides together with oxidation products resulting from the breakdown of components

FUNDAMENTAL RESEARCHES INTO PICTURE CLEANING 69

of the dried linseed oil film. The theoretical basis to this assumption was derived from the known properties of unsaturated triglycerides. In the starting film there is a certain percentage of triglycerides of the saturated kind, formed from palmitic and stearic acid, which either do not participate at all in film polymerization, or do so extremely slowly. Also, oleic acid (having one double bond), being the least reactive of the unsaturated acids, will not readily enter into film formation. The swelling and leaching studies on dried linseed oil films may be used as a guide first to test whether simple triglycerides are leached out of the film (together with oxidation products of lower molecular weight), or whether by using stronger swelling agents it is possible to obtain larger molecular weight species. Thus by employing successively stronger swelling solvents and measuring the leaching obtained in each case one may gain some insight into this. Experimentally it is shown that except for very ineffective swelling solvents (e.g., n-hexane), the leachings are practically the same in quantity and quality irrespective of the degree of swelling. Some of these results are entered in Table 4-6.

TABLE 4-6. LEACHINGS AND DEGREE OF SWELLING

FILM TYPE, MANNER OF AGING	SOLVENT	FILM LEACHING* (%)	DEGREE OF SWELLING ($\Delta V/V_0$)
(i) Stand Oil, containing 0.5% Pb as drier, 200μ thick, aged 7 years 20–30°C. in diffuse light	Methanol	36 ± 3	0.46
	Ethanol	42 ± 3	1.05
	Acetone	45 ± 3	1.32
	Methyl cellosolve	45 ± 3	1.70
	Cellosolve	41 ± 3	1.80
(ii) Barium sulfate/alkali-refined linseed oil; 72.3% pigment, 27.7% oil content, 90μ thick, aged in diffuse light 330 days	n-Hexane	17 ± 1.5	0.07
	Methanol	56 ± 1.5	0.15
	Isopropanol	50 ± 1.5	0.18
	Acetone	52.5 ± 1.5	0.35
	Methylene chloride	46 ± 1.5	1.3
	Chloroform	55 ± 1.5	3.0

*Corrected for organic content of film as in (ii).

Above a certain level of swelling (cf. barium sulfate films), increasing the swelling by using a more active solvent does not appear to remove more soluble material. Swelling with n-hexane is apparently at too low a level to expand the film structure to permit the leaching of the full quantity of soluble material extractable with the higher swelling solvents. There is strong indication, therefore, that a limited molecular weight range of soluble material is being removed; otherwise the percentage of film leaching values would rise with the degree of swelling.

SOLVENT ACTION

By measuring the density of films before and after leaching an estimate of the density of the leached components may be made, provided the percentage of film leaching and the relative composition is known.[31] Calculations made on five types of film, pigmented and unpigmented, show that the density of the leached components is of the order 0.96-1.05, compared with 0.93-0.98 for the starting oils and 1.08-1.15 for the dried virgin films. This also bolsters the hypothesis that low molecular weight species, e.g., triglycerides, are among the substances extracted.

INFRA-RED STUDIES

The leachings of the barium sulfate films (Table 4-6 [ii]) were studied by infra-red spectroscopy. The solutions were evaporated down by blowing with nitrogen at room temperature and the residues were re-dissolved in methylene

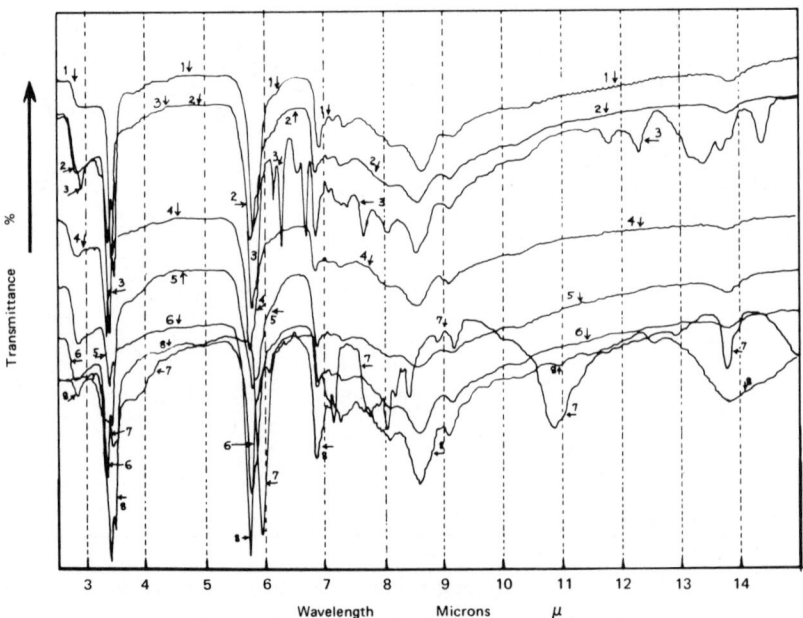

Figure 4-10 The Study of Film Leachings by Infra-red Spectroscopy. Spectrograms 1-6 refer to leachings of $BaSO_4$/linseed oil films: 1—n-hexane; 2—MeOH; 3—iso-PrOH; 4—acetone; 5—methylene chloride; 6—chloroform. Spectrograms 7 and 8 are respectively for azelaic acid and undried linseed oil. The spectrum for azelaic acid was done by the attenuated total reflectance technique. All measurements with Perkin-Elmer Infracord 137.

chloride. Small quantities of these solutions were then evaporated on silver chloride plates and the infra-red spectra measured in the range 2.5-15μ.[32] The resultant spectra are shown in Figure 4-10 for the different extracts. It is seen that there is very little difference between them. The spectra correspond to mixtures of esters with some free acids as indicated by the broadening of the peak, 5.75 to 5.95μ. The spectrum for the original alkali-refined linseed oil is also shown, which indicates the ester peak at 5.75μ but no appreciable acid peak in the vicinity of 5.95μ. The spectrum for pure azelaic acid, for comparison also included in Figure 4-10, shows this peak quite clearly as well as the broadened peak at 3.4μ, evident in the leachings too.

The leachings all show an absorption peak in the 2.9μ region; this is due to the presence of an OH group derived from an acidic component, possibly a hydroperoxide, or even an alcoholic type substance. Again, this 2.9-μ peak is absent in the original linseed oil, which almost entirely is composed of triglycerides and very little free acid or oxidized material.

In Figure 4-11, the infra-red spectra are shown for another experiment in

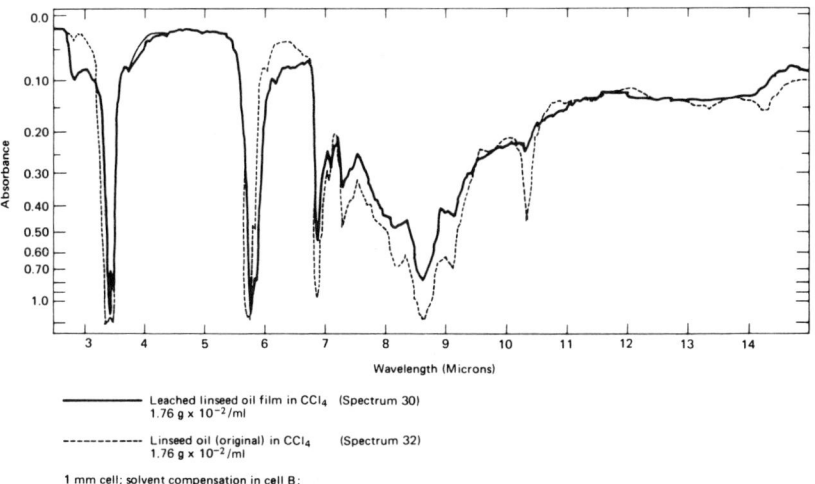

Figure 4-11 *Comparison of Spectra of Film Leachings in Carbon Tetrachloride with Undried Original Oil.*

which a dried linseed oil film was leached with carbon tetrachloride. For comparison the spectrum of the original linseed oil is included. The two spectra were done on equal concentration solutions in infra-red transparent cells. The results are qualitatively the same as before showing the 2.9-μ absorption peak, and the broadened 5.75-μ peak for the leachings.

A more detailed interpretation of infra-red spectra as they apply to linseed oil films may be obtained from the work of Crecelius et al.[33] and also from the special research studies of the Federation of Paint and Varnish Production Clubs.[34] These authors were concerned with the identification of the components of oxidation in whole films, which is of course closely related to the study of leachings. Yamasaki has made a detailed study of the degradation of dried trilinolein films under different conditions and discusses the progressive spectral changes as measured by infra-red spectrophotometry.[35]

On the basis of the infra-red studies undertaken here on the leachings in different solvents, the conclusion may be drawn that the extractable materials are esters—most likely triglycerides, acidic components—such as dicarboxylic acids, and other as yet unidentified oxidation products, which may be hydroperoxides, aldehydes, ketones or even alcohols.

PAPER CHROMATOGRAPHY

The technique of paper chromatography for analyzing the isopropanol leachings of linseed oil films has been applied by P. L. Jones.[36] The films he used were prepared by dipping chromatographic grade paper into a petroleum ether solution of the oil and then permitting them to age for forty-eight weeks under ambient conditions. The percentage leachings for alkali-refined linseed oil and stand oil films of this age, respectively, 37.8 and 16.8%, are of similar order to the present author's values after making due allowance for differences in film age, pigmentation factors, and thickness (Jones' films were approximately 60-μ thick). Jones chromatographed the isopropanol leachings for these films as well as for the films exposed to ultra-violet light for an additional ten weeks. He concluded that essentially the same types of soluble material are extracted, and that the substances from both the stand oil and the alkali-refined linseed oil films are also similar. He concluded that the isopropanol leachings were composed of a heterogeneous mixture of saturated components together with polar material having acidic properties (Jones, p. 122). After saponification of the leachings (and acidification) the chromatograms showed that the free acids were "tentatively identified with stearic and palmitic acid." Presumably these acids were obtained from the saturated esters (triglycerides) of stearic and palmitic acid present in the leachings. Also, his chromatograms after saponification may not have separated the polar material originally present from the saturated esters in the leachings. Polar film breakdown substances, such as the dicarboxylic acids (e.g., azelaic acid), are not easy to identify or separate when also present. Another class of substances suggested by the infra-red spectra studies referred to previously (e.g., hydroperoxides, alcohols) could be missed according to the experimental procedures used. In any case, the work of Jones certainly reinforces the views of the present author on the identity of at least some of the leached components.

GAS CHROMATOGRAPHY

Further clarification of the identity of the leached components is obtained by application of gas chromatography. This method applied to the study of the composition of paint films and their leachings has already been described in some detail by the author.[22] Mills,[37] while essentially studying the properties of whole films, has reported interesting results which corroborate and supplement those facts already known.

Since gas chromatography has become such a useful and productive means for the study of the organic media of works of art, a very brief review of the method as applied to the leaching question is included here. The gas chromatograph consists first of a heated injection assembly where a sample is introduced by a syringe or a hypodermic needle through a rubber septum and is subsequently vaporized. A flow of an inert gas (e.g., nitrogen) flushes the vaporized material out of the injection assembly onto a column. The column of coiled metal or glass contains a carefully conditioned mixture of a stationary phase and an inert phase. Typically, in the analysis of drying oils, the stationary phase is a polyester (e.g., diethylene glycol succinate polyester) which is absorbed onto an inert material such as powdered firebrick. The column is located in an oven where the temperature may be varied. The stationary phase is designed to fractionate the components of the vaporized mixture flowing through the column by the process of adsorption and desorption. Under carefully controlled temperatures and flow conditions, some components will flow through ahead of others owing to less adsorption time spent on the stationary phase of the column. Other components will be more strongly retained by the stationary phase and will appear at the end of the column later. In order to ascertain the appearance of a component at the exit of the column, a detector is necessary. The detector can be of the thermal conductivity type, flame ionization detector, or some other device which can measure the mass of vaporized material arriving at that point in the apparatus. Generally, the detector is designed to give an electronic signal which is transmitted to a recorder. A series of peaks indicates the passage of separated components through the column. A schematic block diagram is shown in Figure 4-12 together with a typical chromatogram. The time of passage of a separated component from the time of injection to detection is referred to as the retention time and the peak area (depending on the type of detector used) can be related to the mass of this component passing through the detector at this time. The shape of the peak itself depends on a number of factors, e.g., manner of injection, separating power of the stationary phase, flow of ni-

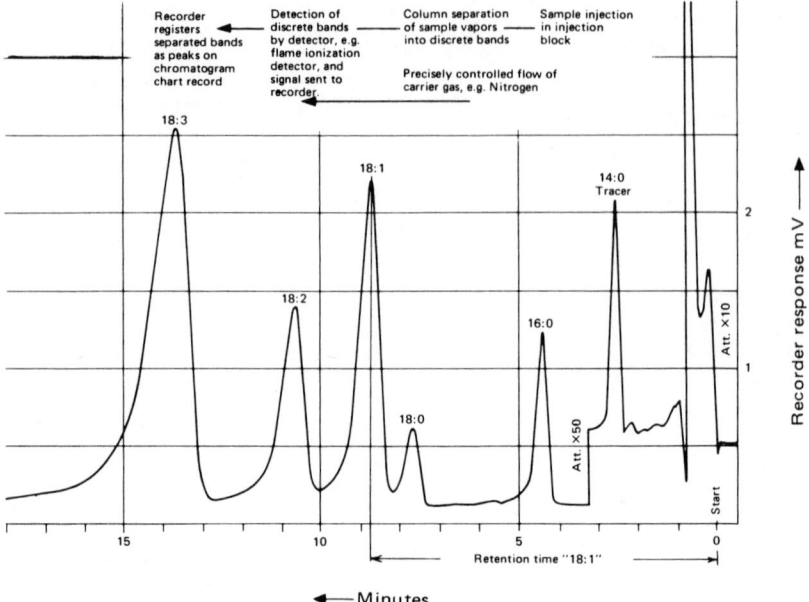

Figure 4-12 Typical Chromatogram of Methyl Esters Derived from Undried Linseed Oil. Column conditions: 2 of 1/8" x 8' EGSS-X (16%) on Gas Chrom P; 175°C., N_2 = 36, Air = 50, H_2 = 20 psi. Methyl esters contain 14:0 methyl ester tracer and are run in CS_2 solution. Perkin-Elmer Gas Chromatograph Model 800.

trogen, temperature, the detector and the stability of its signal to the recorder. It is useful to relate the retention times of the components to one standard component, all considered under the same experimental conditions. The ratio of retention times to a standard component is referred to as the relative retention time for that component. By injecting a series of known components one may arrive at the retention times and the relative retention times for these components which then serve to calibrate the apparatus for the injection of unknowns.

The gas chromatograph successfully employed in this author's laboratory for some years is the Perkin-Elmer Model 800. This apparatus is a dual column instrument with a differential hydrogen flame ionization detector. The two columns are filled with identical material, one is used as a reference column and the other for sample separation. The hydrogen flame ionization detector is exceedingly sensitive to material flowing through and can detect less than 10^{-9} g. under carefully controlled conditions. It is considerably more sensitive than the thermal conductivity detector (katharometer), but suffers from the slight disadvantage that the material going through the detector is combusted or lost. However, by installing a bypass system a known proportion of the column effluent may be trapped out for separate analysis. The hydrogen flame ionization detector will not detect certain substances such as water, hydrogen sulfide, carbon disulfide, carbon tetrachloride, or silicon compounds.

As regards linseed oil films, attention is first focused on the possibility of analyzing the triglycerides present. The technique for this requires generally high temperature columns and is not quite perfected. However, much valuable data can be obtained by converting the triglycerides of the film into the lower boiling and hence more easily separable methyl esters. One such means is by treating the film or leachings with sodium methoxide in methanol solution. Thus the esterification of a triglyceride containing oleic, linolenic and linoleic fragments proceeds as:

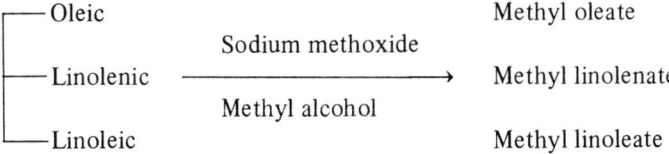

The resulting esters are now sufficiently volatile to be separated on a gas chromatographic column. If some of the components are already in the form of free acids (e.g. palmitic or stearic acid, originally present in an acidic type of linseed oil), or, if dicarboxylic acids, such as those described previously, are present, a modified esterification technique can be used employing boron trifluoride-methanol complex. This converts the triglycerides and free acids to methyl esters. Where it is feasible to convert the triglycerides to free acids first, a saponification technique can be employed, followed by esterification with boron trifluoride-methanol complex. In dried and pigmented films, the same technique can be used; or, where leaching studies are made, solvent extraction of the film followed by esterification can be carried out. The possibility of pigment interference is eliminated here since the esterification is carried out on the solvent leachings.

Table 4-7 lists typical observed relative retention ratios of the methyl ester standards (monocarboxylic and dicarboxylic methyl esters) for the gas chromatographic columns used. This list of standards proved sufficient in assessing a considerable portion of the components in the methyl esters of film leachings.

The leachings of a number of films, pigmented and unpigmented, were esterified and the methyl esters obtained were run in a quantitative fashion

TABLE 4-7. TYPICAL RELATIVE RETENTION RATIOS
FOR VARIOUS METHYL ESTERS ON DEGS AND EGSS-X COLUMNS.
(R.T.R.: RELATIVE TO METHYL STEARATE 1.000)

METHYL ESTER NAME	NO. CARBONS IN ACID PORTION	ABBREVIATION OF METHYL ESTER	MOLECULAR WEIGHT METHYL ESTER	RETENTION TIME RATIO FOR DEGS COLUMN MATERIAL (a)	RETENTION TIME RATIO FOR EGSS-X COLUMN MATERIAL (b)
Monocarboxylic:					
Caproic	6	6:0	130.18	0.100	.....
Enanthic	7	7:0	144.22	0.112	.....
Caprylic	8	8:0	158.24	0.127	.....
Pelargonic	9	9:0	172.27	0.145	.....
Capric	10	10:0	186.30	0.168	0.120
Undecanoic	11	11:0	200.32	0.200	0.160
Lauric	12	12:0	214.34	0.247	0.200
Tridecanoic	13	13:0	228.38	0.300	0.270
Myristic	14	14:0	242.39	0.371	0.350
Pentadecanoic	15	15:0	256.42	0.470	0.450
Palmitic	16	16:0	270.46	0.598	0.600
Heptadecanoic	17	17:0	284.49	0.770	0.780
Stearic	18	18:0	298.51	1.000	1.000
Oleic	18	18:1*	296.50	1.114	1.090
Linoleic	18	18:2*	294.48	1.433	1.330
Linolenic	18	18:3*	292.47	1.860	1.640
Nonadecanoic	19	19:0	312.54	1.300	1.420
Arachidic	20	20:0	326.55	1.711	1.730
Dicarboxylic:					
Oxalic	2	2C2	118.09	0.170	0.100
Malonic	3	2C3	132.12	0.192	0.150
Succinic	4	2C4	146.14	0.252	0.200
Glutaric	5	2C5	160.17	0.315	0.270
Adipic	6	2C6	174.20	0.415	0.380
Pimelic	7	2C7	188.22	0.525	0.510
Suberic	8	2C8	202.25	0.675	0.670
Azelaic	9	2C9	216.28	0.870	0.850
Sebacic	10	2C10	230.31	1.150	1.040

*The abbreviations 18:1, 18:2, 18:3 refer, respectively, to the presence of one, two, or three double bonds of unsaturation. The other methyl esters are saturated.

(a) 15% DEGS on Gas Chrom W at 190°C.; N_2 = 36, Air = 50, H_2 = 20 psi; 2 of $12' \times 1/8''$ columns. F.I.D. (detector).

(b) 16% EGSS-X on Chromosorb P, temperature programmed 160–190°C. at 1° per min.; N_2 = 85, Air = 50, H_2 = 20 psi; 2 of $8' \times 1/8''$ columns. F.I.D. (detector).

DEGS–polyester; diethylene glycol succinate.

EGSS-X–polyester; modified polyester in which the succinic acid molecules have been replaced with organo-silicon molecules.

on the gas chromatograph. The peaks were assigned according to the relative retention ratios (e.g., Table 4-7) of the standards injected into the column under identical experimental conditions. The areas of the peaks were then measured by using a planimeter, or by cutting and weighing of peaks, and the composition of the methyl esters determined. By introducing a known amount of the methyl ester of arachidic acid (20:0) not normally present in these films, the yield could be related to the starting weight of the film (virgin film).

It is assumed that the more saturated triglycerides in the starting linseed oil (or stand oil), containing palmitic (sixteen carbons), stearic (eighteen carbons), and oleic (eighteen carbons with one double bond) do not enter appreciably in film formation but remain essentially unchanged as the film ages. Therefore these acids should be detected upon esterification of the leachings of linseed oil films.

By applying the principle of the random distribution pattern for triglycerides in fatty substances, according to various researchers,[25,26,38] the amounts of the more saturated triglycerides present in the oils of the present researches may be estimated. When this has been done for the alkali-refined linseed oil referred to in a number of film types, it is estimated that, of the *leachable* triglycerides originally present the ratio of palmitic to stearic acids is of the order of 1.6-1.8. Also, the percentage of the more saturated triglycerides (up to four double bonds per triglyceride molecule) of all the triglycerides is of the order of 17%. The analytical data for the alkali-refined linseed oil expressed in these terms is given in Table 4-8.

It is suggested therefore that upon leaching a dried film prepared from this linseed oil, the solubility would be of the order of 17-18%, and that the palmitic plus the stearic portion would amount to about 6%, calculated on the virgin film weight. Also, theoretically, it is suggested that the leachings should contain about 7% oleic acid. In fact the oleic acid content of the leachings would be lower because of oxidation reactions and other film deterioration mechanisms. The observed percentage leaching should actually be of a higher order than 17-18%, since there would also be present a certain amount of oxidation film breakdown products such as dicarboxylic acids, aldehydes, hydroperoxides, and other molecular types. As far as esterification analysis is concerned one could readily determine in the leachings the amounts of palmitic, stearic, oleic and other monocarboxylic acids, as well as the dicarboxylic acids, such as azelaic, suberic, adipic, etc. The other components in the leachings which do not yield methyl esters could very well be the subject of further analytical studies. Also, according to Table 4-8 it would be expected that the leachings should yield the ratio of palmitic acid to stearic acid of about 1.7. This is the calculated ratio whether one considers the triglycerides species soluble up to the level of two, three, or four double bonds per molecule. If it is assumed that, for example, triglycerides up to two double bonds per molecule are leachable, then from the data of Table 4-8 the solubility would be of the order of 3% and the ratio of palmitic to stearic in the leachings still about 1.7. This ratio is the same as in the starting oil and therefore can be used as a sort of internal standard upon which to base the relative amounts of other components present in the leachings. It is expected that the palmitic and stearic components remain unchanged throughout the

TABLE 4-8. CALCULATED TRIGLYCERIDE COMPOSITION
FOR ALKALI-REFINED LINSEED OIL*

TRIGLYCERIDE SPECIES**	NO. DOUBLE BONDS PER TRIGLYCERIDE	PERCENTAGE OF TOTAL TRIGLYCERIDES	PALMITIC 16:0 %	STEARIC 18:0 %	OLEIC 18:1 %
PPP	0	0.03	0.03	— —	— —
SSS	0	0.006	— —	0.006	— —
SPO	1	0.33	0.11	0.11	0.11
SSP	0	0.03	0.01	0.02	— —
SSO	1	0.09	— —	0.06	0.03
SPP	0	0.05	0.035	0.015	— —
PPO	1	0.29	0.20	— — —	0.09
SPL	2	0.23	0.08	0.08	— —
SSL	2	0.07	— —	0.047	— —
SOO	2	0.54	— —	0.18	0.36
PPL	2	0.20	0.13	— — —	— —
POO	2	0.95	0.32	— — —	0.73
OOO	3	1.04	— —	— — —	1.04
SPLe	3	0.81	0.27	0.27	— —
POL	3	1.32	0.43	— — —	0.43
SOL	3	0.75	— —	0.25	0.25
SSLe	3	0.23	— —	0.16	— —
PPLe	3	0.71	0.46	— — —	— —
POLe	4	4.61	1.53	— — —	1.53
SOLe	4	2.61	— —	0.86	0.86
SLL	4	0.26	— —	0.09	0000
OOL	4	2.14	— —	— — —	1.43
PLL	4	0.46	0.16	— — —	— —
Total	0–4	17.76	3.77	2.15	6.90
Remainder	5–9	82.00	3.00	1.65	15.00

*Calculated according to the random distribution pattern for triglyceride drying oils,[25, 26, 38] and for alkali-refined linseed oil. The percentages given are mole%. The composition of the oil by esterification with sodium methoxide in methanol:

palmitic acid	16:0	6.1% by weight
stearic acid	18:0	4.0 by weight
oleic acid	18:1	22.2 by weight
linoleic acid	18:2	15.3 by weight
linolenic acid	18:3	52.5 by weight, Iodine value = 174.5

**The abbreviations refer to P=palmitic (16:0), S=stearic (18:0), O=oleic (18:1), L=linoleic (18:2), and Le=linolenic (18:3). Thus the triglycerides PPP, and SOLe have the configurations:

```
 ┌─P      ┌─S
 ├─P      ├─O
 └─P      └─Le
```

film aging history. However, this may not apply strictly in very severe oxidation atmospheres, where even saturated stable molecules break down.

The leachings from the barium sulfate/alkali-refined linseed oil films referred to in Table 4-4 (viii) had been esterified by the sodium methoxide-methanol technique and analyzed quantitatively by gas chromatography. A typical chromatogram showing the various methyl esters present in the leachings is illustrated in Figure 4-13. The analytical data for the leachings in n-hexane, methanol, isopropanol, acetone, methylene chloride, and chloroform are presented in Table 4-9 in two sections. In part A are given the amounts of components found in the leachings which yield esters; given also are the compositional ratios to the sum of palmitic and stearic methyl esters. In Part B the leaching compositions are related to the oil film content of the paint itself.

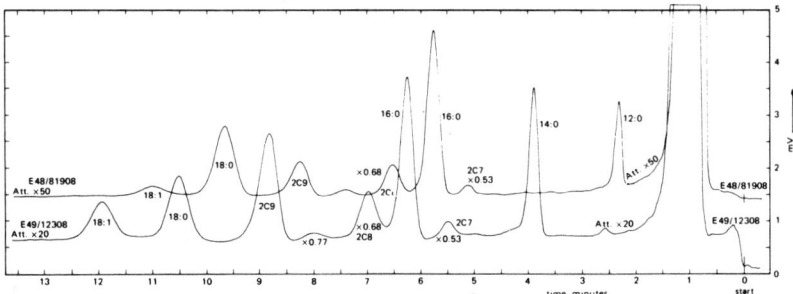

Figure 4-13 Typical Chromatogram of Methyl Esters of Leachings of $BaSO_4$/ Alkali-refined Linseed Oil Films. E48/81908-MeOH leachings esterified by Sodium methoxide technique; E49/12308-MeOH leachings esterified by BF_3-MeOH technique. Column conditions as in Figure 4-12. (12:0 and 14:0 are methyl ester tracers added).

The experimental data in Table 4-9 Part A show the variation in analyzed methyl esters of the leachings in different solvents expressed as a percentage of the organic matter of the virgin paint film. The percentages of material leached (first column) are low for n-hexane (11.2%), and range from 46 to 74% for the other solvents. The other columns give the leached components breakdown compositions in terms of the film. Significantly the leachings in n-hexane are about the same for palmitic, stearic, and somewhat lower for the dicarboxylic acids, compared with the other solvent leachings. The non-esterified portions of the leachings are much lower in quantity for n-hexane. The solvents are tabulated here in order of increasing swelling action: n-hexane to chloroform. Therefore swelling action does not contribute very much to an increase in the extraction by leaching of the palmitic and stearic, but does for the dicarboxylic acids of the film. The differences in the amounts of non-esterified components, 8.5% for n-hexane, compared with 41-70% for the other solvents, suggest that beyond a certain level of swelling (e.g., that for

TABLE 4-9. GAS CHROMATOGRAPHIC ANALYSIS OF METHYL ESTERS OF LEACHINGS OF BARIUM SULFATE/ALKALI-REFINED LINSEED OIL FILMS AND AMOUNTS OF MATERIALS NON-ESTERIFIABLE*

Part A—Analysis of Leachings in Terms of Organic Content

SOLVENT	LEACHINGS AS PERCENTAGE OF ORGANIC FILM CONTENT	METHYL ESTER CONTENT OF LEACHINGS AS PERCENTAGE OF ORGANIC FILM CONTENT			NON-ESTERIFIABLE FRACTION OF LEACHINGS AS PERCENTAGE OF ORGANIC FILM CONTENT
		SUM OF PALMITIC AND STEARIC ESTERS	DICAR-BOXYLIC ESTERS	OTHER ESTERS INCLUDING OLEIC ESTER	
n-Hexane	11.2	2.0	0.2	0.5	8.5
Methanol	74	3.1	1.3	1.3	68
Isopropanol	74	3.0	1.0	0.4	70
Acetone	59	1.9	0.7	0.4	56
Methylene chloride	46	3.8	0.8	0.5	41
Chloroform	56	1.8	0.8	0.5	53

*Barium sulfate/alkali-refined linseed oil—72.3% pigment, 27.7% oil content in dried film 90-μ thick, aged in diffuse light 300–330 days. The esterifications of the leachings were carried out by the sodium methoxide technique. The peaks were measured and compared on a weight of component basis. The flame ionization detector was calibrated as to its sensitivity to each peak component considered, and for calculation of the quantities of methyl esters, either lauric or arachidic methyl ester was used as an internal standard.

Part B—Analysis of Methyl Esters of Leachings

SOLVENT	RATIOS OF COMPONENT METHYL ESTERS TO THE SUM OF THE PALMITIC AND STEARIC METHYL ESTERS									
	GLUTARIC	ADIPIC	PIMELIC	PALMITIC	SUBERIC	AZELAIC	STEARIC	OLEIC	LINOLEIC	LINOLENIC
n-Hexane (i)	—	0.02	0.02	0.59	0.02	0.03	0.41	0.21	—	—
n-Hexane (ii)	—	—	0.02	0.61	0.05	0.03	0.39	0.05	0.03	—
Methanol (i)	—	—	0.05	0.62	0.13	0.24	0.38	0.09	0.01	—
Methanol (ii)	—	—	0.04	0.59	0.14	0.24	0.41	0.08	—	—
Isopropanol (i)	—	—	0.02	0.60	0.10	0.18	0.40	0.19	—	—
Isopropanol (ii)	—	—	0.04	0.61	0.15	0.22	0.39	0.13	—	—
Isopropanol (iii)	—	—	0.03	0.59	0.12	0.20	0.41	0.10	—	—
Acetone	—	—	0.03	0.57	0.15	0.30	0.43	0.12	—	—
Methylene chloride (i)	—	—	—	0.56	0.05	0.18	0.44	0.14	—	—
Methylene chloride (ii)	—	—	—	0.59	0.06	0.14	0.41	0.17	—	—
Chloroform	—	—	0.06	0.59	0.20	0.32	0.41	0.12	0.02	—
Components of original alkali-refined linseed oil	—	—	—	0.61	—	—	0.39	2.19	1.51	5.20

methanol) the film structure is sufficiently opened up to permit other molecular components to be extracted.

In Part B the methyl esters of the leachings were analyzed and, significantly, the ratios of palmitic and stearic are quite constant and, within experimental error, identical with that of the original oil. Of the dicarboxylic acids the main components identified are pimelic, suberic, and azelaic. The leachings in n-hexane give significantly lesser quantities of azelaic, and again this appears to be due to the lower state of swelling of the film in that solvent. The other components linoleic and linolenic are present in very small quantities, not measurable under the experimental conditions, and presumably have been used up in film formation or oxidation-degradation reactions.

Referring again to Table 4-9 Part A, the leaching action of solvents (methanol ... to chloroform) results in the extraction of amounts of the more saturated triglycerides which give upon analysis about 2-4% of the saturated acid esters palmitic plus stearic. If one makes allowance for incomplete leaching time or lack of quantitative yield in the method of conversion of the triglycerides to methyl esters, then the results of 2-4% compare favorably with the amount of 6% calculated from the random distribution theory (cf. Table 4-8 where the triglycerides having 0-4 double bonds—C=C—produce about 6% of palmitic and stearic combined). These experiments so far permit the conclusions to be drawn that: a major portion of the unreacted and more saturated triglycerides can be readily extracted in this molecular form from the film, even with such low swelling solvents as n-hexane; and, that other components (e.g. dicarboxylic acids and non-esterifiable products, also possibly low molecular weight polymers and other fragments) are extractable according to a certain threshold amount of swelling. Here such factors as polarity and affinity with the solvent in question may enter, along with diffusion characteristics.

Table 4-10 includes the results of gas chromatographic analysis of the leachings of a variety of films of different age, and also of a group of films aged under accelerated conditions using strong ultra-violet lamps. The method of esterification generally used here is based on boron trifluoride in methanol. This method is more efficient than the sodium methoxide method in the conversion of dicarboxylic acids to their corresponding methyl esters. Thus the quantities given in the column heading "dicarboxylics" are generally higher than those given in the corresponding column of Table 4-9-A, where comparisons can be made. Some significant conclusions may be drawn from the data given in Table 4-10. It will be seen that generally the palmitic plus the stearic methyl esters (expressed as a per cent of the film organic content) are in the range of 4-6%, which corresponds fairly closely with the expected amount of 6% calculated from the random distribution theory. This is true, independently of pigment content and of whether the aging is under normal conditions or under ultra-violet light. The next significant finding is that the extracted dicarboxylics are relatively greater in the films aged under ultra-violet light than in those aged under normal conditions. This is especially true in the films based on iron oxide, titanium dioxide (rutile), and white lead. This can perhaps be explained on the basis of the high absorption these

TABLE 4-10. GAS CHROMATOGRAPHIC ANALYSIS OF METHYL ESTERS AND AMOUNTS OF NON-ESTERIFIABLE SUBSTANCES IN THE LEACHINGS OF VARIOUS LINSEED OIL AND PIGMENTED FILMS

FILM TYPE AND MANNER OF AGING	LEACHING SOLVENT	LEACHING AS PERCENTAGE OF ORGANIC CONTENT OF FILM	METHOD OF ESTERIFICATION	METHYL ESTER CONTENT AS PERCENTAGE OF ORGANIC CONTENT OF FILM			NON-ESTERIFIABLE PORTION OF LEACHINGS AS PERCENTAGE OF ORGANIC CONTENT OF FILM
				PALMITIC AND STEARIC	DICARBOXYLICS	OTHER, INCLUDING OLEIC	
A. Natural Aging in Diffuse Light; 20–25°C, 40–45% R.H.:							
(i) Barium sulfate/alkali-refined linseed oil; 72.3% pigment, 27.7% oil; 90μ thick, aged 330 days	Methylene chloride & methanol	46	BF_3-MeOH,* HCl-MeCl$_2$	4.8	7.0	2.2	31
(ii) White lead/alkali-refined linseed oil; 84.3% pigment, 15.7% oil; 120μ thick, aged 330 days	Methylene chloride & methanol	26	BF_3-MeOH, HCl-MeCl$_2$	4.8	4.0	5.0	12
(iii) Iron oxide/alkali-refined linseed oil; 80.5% pigment, 19.5% oil; 76μ thick, aged 410 days	Methylene chloride & methanol	55	BF_3-MeOH, HCl-MeCl$_2$	3.7	6.4	2.9	42

(iv) Titanium dioxide (rutile)/alkali-refined linseed oil; 71.9% pigment, 28.1% oil; 90μ thick, aged 400 days	Methylene chloride & methanol	60	BF$_3$-MeOH, HCl-MeCl$_2$	4.3	7.0	2.0	47
(v) Alkali-refined linseed oil, no pigment; 35μ thick, aged 340 days	Methylene chloride & methanol	65	BF$_3$-MeOH, HCl-MeCl$_2$	4.5	9.5	5.0	46
(vi) Alkali-refined linseed oil, no pigment; 35μ thick, aged 340 days	Methanol	65	NaOMe**	4.9	1.7	1.9	56
(vii) White lead/alkali-refined linseed oil; 77% pigment, 23% oil; 100μ thick, aged 14 years	Ethylene dichloride	55	NaOH-MeOH, BF$_3$-MeOH	5.7	4.0	1.3	44
(viii) White lead/stand oil; 81% pigment, 19% oil; 130μ thick, aged 15 years	Acetone	22	BF$_3$-MeOH†	4.0	3.5	1.9	13
(ix) White lead/stand oil; 81% pigment, 19% oil; 130μ thick, aged 15 years	Methylene chloride	26	BF$_3$-MeOH	3.3	1.6	0.9	20

B. ††Accelerated Aging Under Strong Ultra-violet Light, 20°C. 50% R.H.:

(x) Barium sulfate/alkali-refined linseed oil; 77.5% pigment, 22.5% oil; 100μ thick, aged 909 days	Methylene chloride & methanol	27	BF$_3$-MeOH	5.5	8.8	1.2	11
(xi) White lead/alkali-refined linseed oil; 84.7% pigment, 15.3% oil; 100μ thick, aged 909 days	Methylene chloride & methanol	26	BF$_3$-MeOH	5.5	11.0	1.3	8

FILM TYPE AND MANNER OF AGING	LEACHING SOLVENT	LEACHING AS PERCENTAGE OF ORGANIC CONTENT OF FILM	METHOD OF ESTERIFICATION	METHYL ESTER CONTENT AS PERCENTAGE OF ORGANIC CONTENT OF FILM			NON-ESTERIFIABLE PORTION OF LEACHINGS AS PERCENTAGE OF ORGANIC CONTENT OF FILM
				PALMITIC AND STEARIC	DICARBOXYLICS	OTHER, INCLUDING OLEIC	
(xii) Iron oxide/alkali-refined linseed oil; 85% pigment, 15% oil; 100μ thick, aged 909 days	Methylene chloride & methanol	85[a]	BF$_3$-MeOH	6.0	21.0	2.0	56
(xiii) Titanium dioxide (rutile)/alkali-refined linseed oil; 75% pigment, 25% oil; 100μ thick, aged 909 days	Methylene chloride & methanol	84[a]	BF$_3$-MeOH	5.0	14.0	1.0	64
(xiv) Alkali-refined linseed oil, no pigment; 50μ thick, aged 909 days	Methylene chloride & methanol	27	BF$_3$-MeOH	3.0	4.0	1.5	18.5

C. Naturally Aged Films from Old Paintings

Sample					
(xv) Iron oxide/linseed oil ground layer from 17th-century Dutch master painting. Estimated 80% pigment, 20% oil content.	Methylene chloride & methanol	—	BF$_3$-MeOH	5^b	15^b	1^b
(xvi) White lead/linseed oil paint fragment from 17th-century Dutch master painting. Estimated 80% pigment, 20% oil content.	Methylene chloride & methanol	—	BF$_3$-MeOH	7^b	10^b	—

*BF$_3$-MeOH, HCl-MeCl$_2$ refers to the esterification with boron trifluoride-methanol with the addition of a small quantity of hydrochloric acid and some methylene chloride.

**The sodium methoxide (NaOMe) in methanol esterification technique, which is somewhat inefficient in esterifying dicarboxylic acids—but highly efficient in esterifying triglycerides by transesterification.

†The BF$_3$-MeOH technique gave approximately the same degree of esterification as that employing HCl and MeCl$_2$ (as additional reaction components).

††There were two ultra-violet lamps of the Hanovia type used: 654A, 200 W., 125 V., high-pressure quartz mercury vapor lamps supplying energy in the filtered ultra-violet region of the order of 2.5 X 10^3 ergs per sec per cm.^{2}, and 2.8 X 10^3 ergs per sec per cm.^{2} in the visible at the sample surfaces.

^{a}The films disintegrated very quickly upon solvent contact with dispersion of pigment.
^{b}The yield of methyl esters could not be calculated owing to the very small size of these samples. The figures given may be in error by 20%

pigments have for ultra-violet light, which promotes in turn more oxidative reactions within the film structure. Ultra-violet light exposure is known to promote the formation of dicarboxylic acids along with aldehydes, ketones, and other oxygen-containing molecules.[33,35] The exposure of the films for 909 days profoundly increased the solubility by leaching action of the iron oxide and titanium dioxide (rutile) films to 85 and 84%, respectively. These films almost disintegrated at once upon contact with solvent thereby indicating that they were in a very advanced state of deterioration.

The ultra-violet aging of the white lead films apparently had little effect upon the percentage of leaching, although extreme film brittleness developed. The barium sulfate films also became greatly embrittled and less soluble in solvent. These solubility differences were expressed almost entirely in the increase (or decrease) of the relative amounts of non-esterifiable material. Again it would be worthwhile to examine the nature of this significant portion of the leachings.

No attempt was made to assign an "accelerated time factor" to the ultra-violet-exposed films compared with those aged under normal conditions. Presumably the ultra-violet exposure accelerates oxidation reactions as shown in increasing amounts of dicarboxylic acids in the leachings. The kind of brittleness resulting from this sort of aging is similar to that observed in paint films of great age. Perhaps one year under the accelerated conditions corresponds to 20-30 years of natural aging, all other factors being equal.

In Section C of Table 4-10 some fragmentary results are included for paint films taken from an important seventeenth-century Dutch painting. Owing to the smallness of the sample, approximately 1 mgm, the pigment-oil content proportions could not be ascertained with any reasonable accuracy, and hence the amount of leaching could not be ascertained. Typical amounts of palmitic, stearic, and dicarboxylic acids were obtained showing the same type of leaching mechanism as with younger films. However, the iron oxide sample appeared to have relatively more dicarboxylic acids. It should be mentioned that the samples here were removed from interior layers of the painting to avoid the possibility of varnish or other contaminations. A special technique was developed whereby a small portion of canvas was scraped away from the reverse of the painting and then the relevant paint layer was sampled. As a guide in this a preliminary cross section was made to determine the layer structure, thickness of layers, and pigment composition.

In an earlier section it was stated that the amount of leaching action depends upon the duration of contact with solvent (cf. Figure 4-8). It would therefore be of interest to analyze the leachings at different times in the early period of solvent contact. Some preliminary experiments in this regard are worth reporting.

An unpigmented alkali-refined linseed oil film, aged for three years under ultra-violet light, was leached in chloroform in two stages of four hours each, and each extract was analyzed by gas chromatography of the methyl esters. As the results of Table 4-11 show, the greater bulk of the non-esterifiable portion and major portions of the palmitic, stearic, and dicarboxylic components are removed in the first leaching. The second leaching results in progressive removal of more of the palmitics, stearics, and dicarboxylics, and at this point very little of the non-esterifiable portion is extracted. This indicates that the non-esterifiable substances are extracted relatively easily from the film structure compared with the leachable triglycerides and the dicarboxylic acids. Of interest too is the study of the change of composition of leachings from a film that has already been leached but has been allowed to undergo further aging before the second solvent treatment. Previously mentioned studies (cf. Table 4-5) of the influence of aging upon the total amount of leaching suggested that, as the film ages, additional amounts of soluble material may be extracted.

TABLE 4-11. EFFECT OF TIME OF LEACHING UPON THE COMPOSITION OF THE LEACHED COMPONENTS AS DETERMINED BY GAS CHROMATOGRAPHY*

	LEACHINGS AS PERCENTAGE OF VIRGIN FILM	METHYL ESTERS AS PERCENTAGE OF VIRGIN FILM			NON-ESTERIFIABLE PORTION OF LEACHINGS AS PERCENTAGE OF VIRGIN FILM
		PALMITIC AND STEARIC	DICARBOXYLICS	OTHER, INCLUDING OLEIC	
First 4 hrs.	13.7	1.0	1.4	0.5	10.8
Second 4 hrs.	2.3	0.3	0.5	0.3	1.2
Totals.	16.0	1.3	1.9	0.8	12.0

* Specimen material: Alkali-refined linseed oil, 50μ thick.
Specimen preparation: Aged three years under strong ultra-violet light (see Note, Table 4-10).
Leaching solvent: Chloroform.
Esterification method: Boron trifluoride-methanol.

The results of Table 4-12 (white lead/stand oil films) show that further aging results in further extractable amounts of palmitic, stearic, and dicarboxylics, but significantly smaller amounts of the non-esterifiable components.

TABLE 4-12. EFFECT OF AGING OF A LEACHED FILM UPON THE COMPOSITION OF FURTHER LEACHINGS AS DETERMINED BY GAS CHROMATOGRAPHY*

	LEACHINGS AS PERCENTAGE OF VIRGIN FILM	METHYL ESTERS AS PERCENTAGE OF VIRGIN FILM			NON-ESTERIFIABLE PORTION OF LEACHINGS AS PERCENTAGE OF VIRGIN FILM
		PALMITIC AND STEARIC	DICARBOXYLICS	OTHER, INCLUDING OLEIC	
(a) Acetone, 15 yrs.	22	4.1	3.5	2.4	12.5
(b) Methylene chloride, 15 yrs.	26	3.3	1.6	0.9	20.2
(c) 8 yrs; 7 yrs; acetone	5.3	2.1	0.9	0.1	2.2
(d) 8 yrs; 7 yrs; methylene chloride	4.3	1.5	0.5	0.2	2.2

*Specimen material: All films were white lead/stand oil; 81% pigment, 19% oil; 130μ thick. All films were aged under normal conditions.

Specimen preparation:

(a) Aged fifteen years, then acetone-leached for the first time. Leachings chromatographed.
(b) Aged fifteen years, then methylene chloride-leached for the first time. Leachings chromatographed.
(c) Aged eight years, acetone-leached, then the leached film was aged seven years more and leached a second time in acetone. Second leachings chromatographed.
(d) Aged eight years, acetone-leached, then the leached film was aged seven years more and leached a second time in methylene chloride. The second leachings were chromatographed.

All esterifications by the boron trifluoride-methanol method.

SWELLING ACTION IN DIFFERENT SOLVENTS

Attention will now be given to the capacity of leached films to swell in solvents. The unique swelling behavior of virgin films was described earlier, and it was noted that contact with solvent for a second time results in a swelling curve which rises quickly at first and steadily reaches an equilibrium maximum value (cf. Figure 4-7). This phenomenon of swelling is generally reproducible in leached films, provided the leaching is complete for the solvent in question and that no further aging of the film structure has taken place. Under certain circumstances, however, the degree of swelling may be altered by replacing the swelling solvent still imbibed in the film with another one.

MOLECULAR VOLUME OF SOLVENTS

The first thought that occurs on surveying the change in swelling in different solvents is that perhaps the molecular volume has some effect. Thus in Figure 4-14 the different amounts of swelling are plotted against the molecular volume of the solvent. Described here is the behavior of a white lead paint film which has been pre-leached in methyl alcohol. This film contains 19% stand oil medium in the virgin film, has been aged for forty-six weeks at 32°C. and is 200-μ thick. The vertical scale represents the equilibrium degree of swelling, i.e., the volume of solvent taken up by unit volume of film under equilibrium conditions. The horizontal scale refers to the molecular volume of the solvent (a measure of the volume of the molecule of the solvent). This does not, however, take into account the shape of the solvent molecule which is of some importance. Nevertheless it is interesting to relate the swelling to the molecular volume.

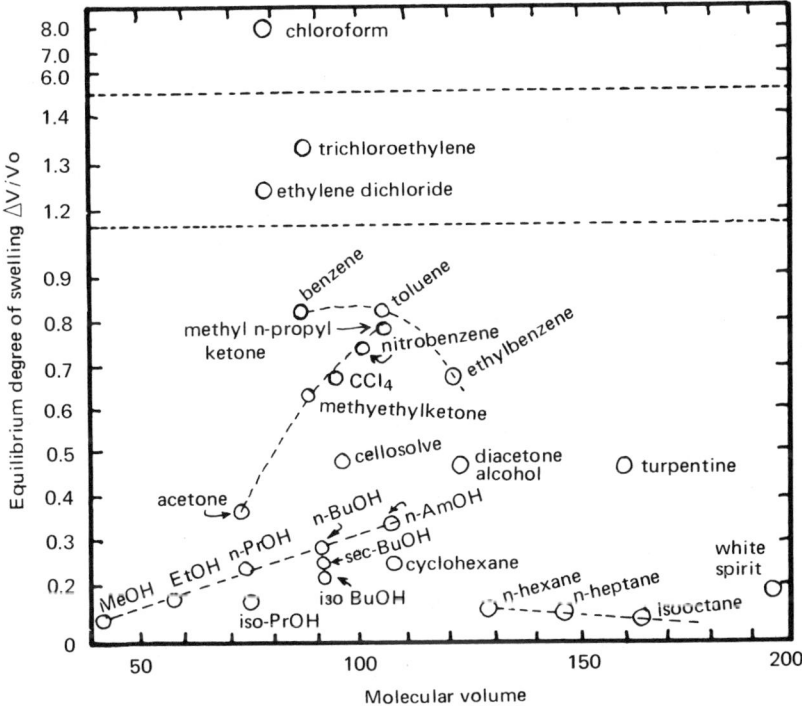

Figure 4-14 The Swelling of Methanol-leached White Lead/Stand Oil Films in Various Solvents at 22°C. in Terms of the Molecular Volume of Solvent. Films aged forty-six weeks at 32°C., 200μ thick before methanol leaching.

Consistent behavior is observed with the normal alcohols where the degree of swelling increases with the increasing molecular size of the alcohol. The ketones also exhibit this behavior, while the aromatic hydrocarbons behave in a reverse fashion. By way of contrast, the paraffinic hydrocarbons, hexane, heptane, isooctane, have no appreciable effect on the degree of swelling. Noteworthy is the very high degree of swelling obtained with each of the chlorinated hydrocarbons, especially chloroform. Returning to the alcohols, it will be observed that the shape factor does determine the amount of swelling for molecules of the same molecular volume. Thus normal butyl alcohol swells films more than isobutyl alcohol, and normal propyl alcohol swells more than isopropyl alcohol.

SOLUBILITY PARAMETERS

Another approach to the interpretation of the diverse swelling results is by means of the thermodynamic theory of polymer-solvent solutions, in particular the "cohesive energy density" concept introduced by Hildebrand and his co-workers.[39] By cohesive energy density is meant the amount of energy per cubic centimeter of solvent, or polymeric substance, required to overcome completely all the internal forces that keep the molecules of the substance together. It can readily be seen how this term is related to solubility behavior. When a substance dissolves, the molecules of the dissolving substance are dispersed by the solvent. This happens in the case of straightforward solution. However, when an amorphous polymeric material, such as dried linseed oil, is treated with solvent, swelling results. In order to disperse such a substance into solution, it would be necessary to rupture completely the three-dimensional structure. The condition of complete solubility, although not attainable in practice, may be approached with certain solvents rather than others. The solvents having the greatest swelling action on the dried linseed oil film are those in which the cohesive energy densities of the polymeric film and the solvent are practically equal. In such instances the attraction of the polymer and solvent molecules are at the greatest level and, were it not for the special structure keeping the film intact, complete solubility would result.

In practice reference is made to the square root of the cohesive energy density, often referred to as the solubility parameter. Each solvent and, where it can be calculated, each polymeric substance, has a solubility parameter number associated with it. The solubility parameter of a polymeric substance such as dried linseed oil may be obtained by examining its swelling behavior in a variety of solvents. The parameter is equal to that of the solvent which causes maximum swelling.

It is not possible to discuss here in detail the theory behind the solubility parameter concept which has adequately been dealt with by Burrell.[40,41]

The solubility parameters of a considerable number of solvents have been determined by various researchers and are listed by Burrell and Gardon.[41,42] These parameters, symbolized as δ_s, range numerically from about 5 for fluorocarbons to 23.2 for water. The solubility parameters of those solvents studied in leaching and swelling experiments are listed separately in Table 4-13. (They are also given in Appendix F, together with other physical characteristics.)

TABLE 4-13. SOLUBILITY PARAMETERS δ_s OF SELECTED SOLVENTS

Solvent	δ_s	Solvent	δ_s
Varsol	7.0	Benzene	9.2
Shell sol 715	7.0	Diacetone alcohol	9.2
Petroleum ether	7.0	Chloroform	9.3
Isooctane	7.3	Methyl ethyl ketone	9.3
n-Hexane	7.3	Trichloroethylene	9.3
n-Heptane	7.4	Tetrachloroethylene	9.4
Diethyl ether	7.4	Methyl acetate	9.6
V. M. & P. naphtha	7.6	Methylene chloride	9.7
Kerosene	7.6	Ethylene dichloride	9.8
White spirit (BS 245/56)	7.6	Tetrahydrofuran	9.9
Cyclohexane	8.2	Cellosolve	9.9
Methyl isobutyl ketone	8.4	Cyclohexanone	9.9
Dipentene	8.5	Acetone	10.0
n-Amyl acetate	8.5	1,4-dioxane	10.0
n-Butyl acetate	8.5	n-Octanol	10.5
Shell cyclosol 53	8.5	Pyridine	10.7
Turpentine	8.5	Methyl cellosolve	10.8
1,1,2-Trichloroethane	8.5	n-Amyl alcohol	10.9
Diethyl benzene	8.7	Isobutyl alcohol	11.1
Di-isopropyl benzene	8.7	n-Butyl alcohol	11.4
Cellosolve acetate	8.7	Cyclohexanol	11.4
Methyl n-propyl alcohol	8.7	Isopropyl alcohol	11.5
Ethyl benzene	8.8	n-propyl alcohol	11.9
n-Propyl acetate	8.8	Dimethyl formamide	12.1
Xylene (o,m,p)	8.8	Ethanol	12.7
Toluene	8.9	Methanol	14.5
n-Butyl cellosolve	8.9	Glycerol	16.5
Ethyl acetate	9.1	Water	23.2

In Figure 4-15 (a, b, and c) are plotted the equilibrium degrees of swelling for leached white lead/stand oil films, twenty-seven weeks, seven years, and fourteen years aging, as a function of the solubility parameter δ_s. The swelling was measured on free films employing the dimensional measurement technique as well as the weighing method. The film fragments were evacuated each time to expel all residues of solvent before testing a new solvent. Thus each point on the graphs of Figure 4-15 represents an individual swelling experiment followed by complete evaporation of solvent.

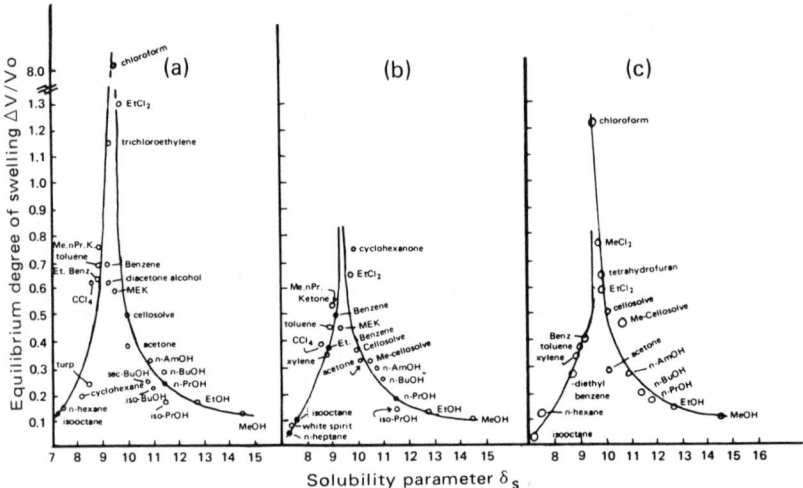

Figure 4-15 The Equilibrium Degree of Swelling at 22°C. in Terms of Solubility Parameters of Solvents for Leached White Lead/Stand Oil Films Aged at 20-30°C.: (a) 27 weeks (b) 7 years (c) 14 years.

It will be observed that a characteristic curve is obtained which rises sharply in the region of solubility parameters 9-10 units. The actual maximum is estimated to be 9.3-9.5 units, which corresponds to maximum swelling of the film in question. Thus the solubility parameter of the film (stand oil) is said to be of the value 9.3-9.5 units. The strongly swelling solvents, such as ethylene chloride, trichloroethylene, methylene chloride, chloroform, tetrahydrofuran, or cyclohexanone, all have solubility parameters in the range 9.3-9.9 units. Solvents with solubility parameters well removed from the "peak" of the curve, i.e., in the region of 7-8 or greater than 13 units, have little swelling effect on the linseed oil film structure. Thus the petroleum solvents and alcohols, such as ethyl and methyl alcohol, are low swellers. (It should be borne in mind here that a low sweller is not necessarily a poor leaching solvent. The swelling curves discussed here correspond to leached films, and for leached linseed oil film structures methanol is a low sweller.) Solvents closer in value to the central region, such as acetone, cellosolve, benzene, or carbon tetrachloride, have moderate to extensive swelling action. Similar behavior was noted in alkali-refined linseed oil films and in unpigmented films.

It is interesting to note that the effect of film aging is to depress the shape of the curve—compare (a) and (c)—but this does not alter the general relationship among the solvents. The effect of aging is shown more graphically in Figure 4-16.

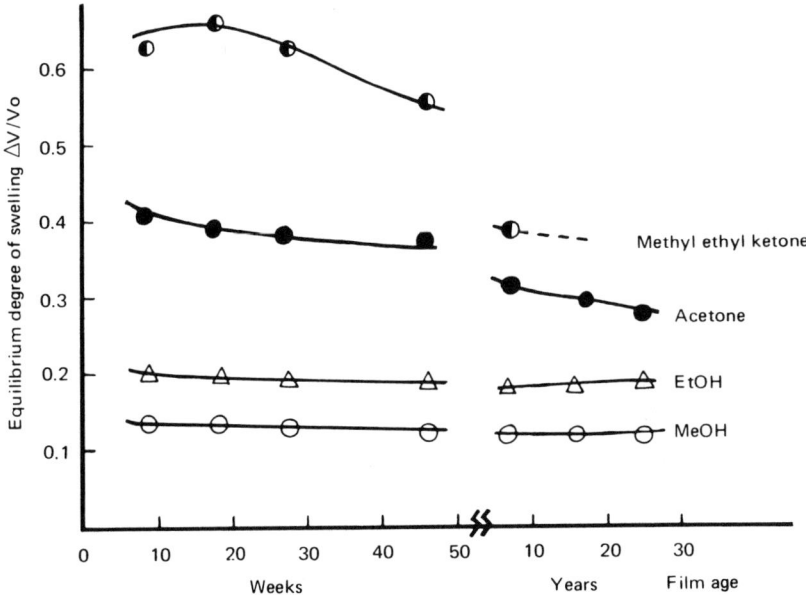

Figure 4-16 The Effect of Film Age on the Equilibrium Degree of Swelling. Pre-leached white lead/stand oil films—swollen at 22°C.

From the solubility parameter-swelling curves it should be possible to anticipate the behavior of a new solvent. One simply locates the value of the solubility parameter for that solvent on the curve. It should be mentioned that in more recent theoretical discussions on solubility parameters certain correction factors pertaining to internal solvent characteristics (hydrogen bonding, polarity, etc.) have been recommended.[42] As a good approximation, the uncorrected solubility parameter figures given in Table 4-13 are sufficient for present purposes.

It is of interest to consider the effect of solvent mixtures on the swelling of linseed oil films. According to the above theory, it can be predicted that solvents chosen from different sides of the maximum in the solubility parameter curve will produce enhanced swelling action when mixed together. In Figure 4-17, mixing benzene and methyl alcohol results in maximum swelling. On the other hand, mixing two solvents which have solubility parameters on the same side of the region 9.3-9.5 units, e.g., benzene and n-hexane, produces more or less linear behavior without a maximum in the curve. Thus the solubility parameters of mixtures of solvents are additive according to their relative amounts.

A series of swelling measurement experiments were carried out in which a swollen film was transferred immediately to a solvent of lower or higher

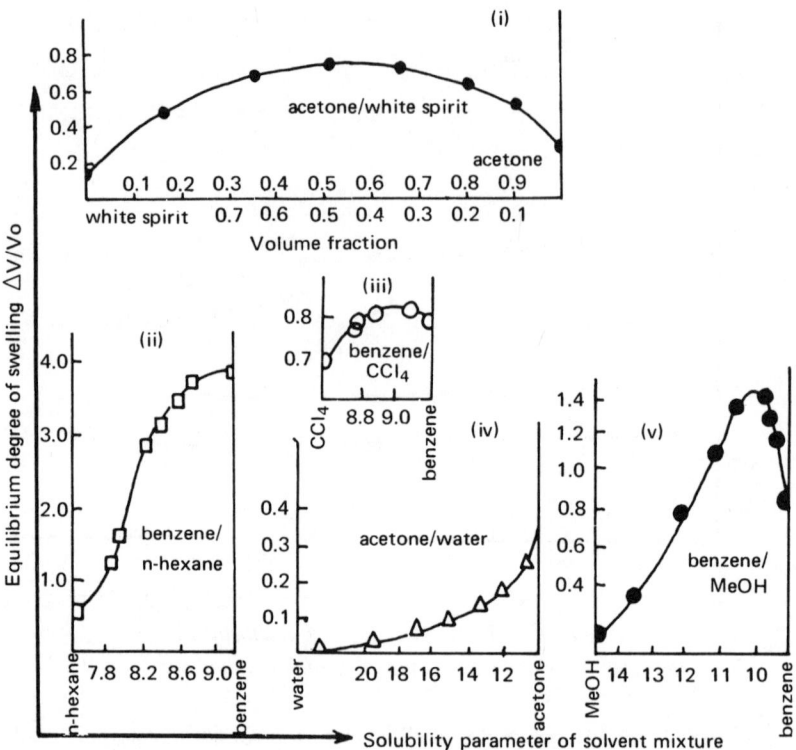

Figure 4-17 The Solubility Parameter of Solvent Mixtures and the Equilibrium Degree of Swelling. (i), (iii), (iv), (v): leached white lead/stand oil films, aged for forty-six weeks at 32°C.–swelling at 22°C.; (ii): leached vacuum stand oil, aged for 100 days 65°C.

swelling potential. It was observed that abnormal degrees of swelling could be obtained, but that after solvent evaporation the film structure would swell in the normal manner. Some typical results on white lead/stand oil films are given in Table 4-14.

The effect of substituting a high swelling solvent (e.g. cyclohexanone with acetone) results in increased levels of swelling. Thus the normal value for acetone is 0.31, and by pre-swelling with cyclohexanone the value of 0.35 is obtained. Thus a high swelling solvent can expand the film structure and be replaced by a low sweller. The film will not shrink back completely. On the other hand, if a low swelling solvent were first applied (e.g., the petroleum solvent Shell sol 715) followed by placing the swollen film in acetone, the film will swell to the level expected for acetone. In both sets of experiments,

TABLE 4-14.* EFFECT OF SOLVENT SUBSTITUTION ON
THE EQUILIBRIUM DEGREE OF SWELLING—$\Delta V/V_0$

SOLVENT EXPERIMENT	$\Delta V/V_0$	SOLVENT EXPERIMENT	$\Delta V/V_0$
(i) Swelling in acetone and evaporation of solvent	0.31	(i) Swelling in acetone and evaporation of solvent	0.31
(ii) Swelling in cyclohexanone	0.76	(ii) Swelling in Shell sol 715	0.06
(iii) Immediate transfer of swollen film to acetone and evaporation of solvent	0.35	(iii) Immediate transfer of swollen film to acetone and evaporation of solvent	0.31
(iv) Re-swelling in acetone	0.31	(iv) Re-swelling in acetone	0.31

*White lead/stand oil films aged seven years under normal conditions. The values of $\Delta V/V_0$ are for equilibrium and are determined on unsupported films by dimensional measurements.

upon evaporation of the solvents and re-swelling in acetone, the normal film swelling behavior was regained. This type of swelling behavior is of course well known with other gels, such as gelatine, rubber, cross-linked polymers, and so forth. In another set of experiments paint films were progressively immersed in a series of alcohols of increasing swelling power, and then the experiments repeated by progressive immersion in the same series of alcohols but in descending swelling power. It was observed that in the ascending swelling series normal values were obtained for the degree of swelling, but higher values were obtained in the descending series.

THE RATE OF SWELLING ACTION; DIFFUSION OF SOLVENTS

In the previous section consideration was given to the amounts of swelling in solvents under equilibrium conditions. It is very important to know how fast these solvents can penetrate the structure of linseed oil films and to discover what natural laws apply. On the basis of extensive studies of the rate of swelling of such films it was found that the theory of diffusion, as represented by Fick's law, applied fairly well. A good discussion of the theory of diffusion as applied to various materials is given by Barrer[43] and Crank.[44] Fick's law of diffusion may be expressed in a modified form, as suggested by Prager and Long,[45] Crank,[46] and others,[47] at least for the initial stages of diffusion as:

$$\frac{Q}{Q\infty} = \frac{K\sqrt{t}}{l}$$

where Q and $Q\infty$ refer to the amount of sorption (or swelling) at time t and at equilibrium (i.e., infinite time), l is the thickness of the film in question, and K is a constant term which incorporates the diffusion constant for the film and vapor (solvent) in question. According to this relationship, a graphical plot of the ratio $Q/Q\infty$ versus the square root of the time should give a straight line from which the diffusion coefficients or diffusion behavior of the solvents can be measured and compared.

Experimentally, the rate of swelling was carried out on leached supported films using the swelling measurement apparatus previously described. Here one dimension, the thickness, could readily be followed with time. The evaporation rate of solvents from films could also be determined in this way or, with free films, by measuring dimensional changes or by following changes in weight.

In Figure 4-18 swelling rates are assessed for different films and of different thicknesses by plotting the fraction of total swelling reached ($Q/Q\infty$) as a function of time (t).

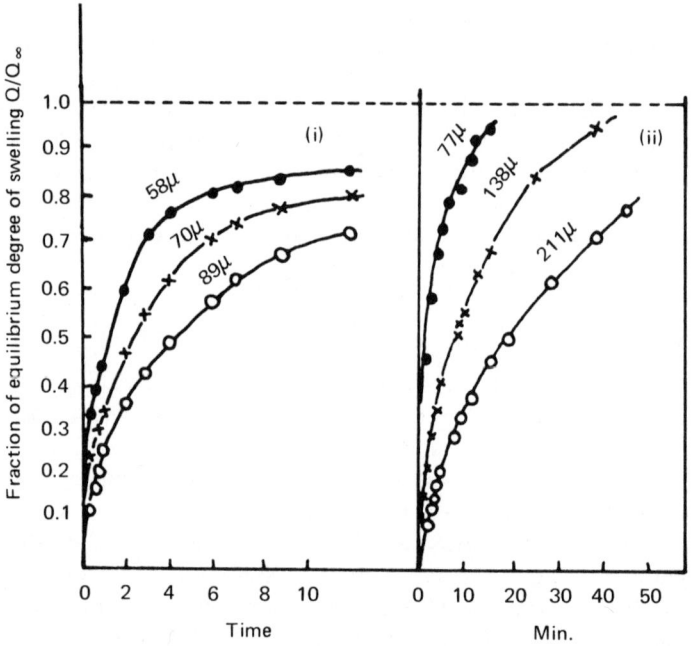

Figure 4-18 The Effect of Film Thickness on the Rate of Swelling at 22°C. (i) Leached stand oil, aged three weeks at 32°C. supported on glass, swelling in CCl_4. (ii) Leached white lead/stand oil, aged seventeen weeks at 32°C., supported on glass, swelling in MeOH.

It is readily seen that the film thickness has a profound effect on the time taken for a given amount of swelling. Thus if one considers the times for half of the maximum amount of swelling to take place, (i.e., $Q/Q\infty = 0.5$) then the graphs show that as the thickness is doubled the corresponding times are quadrupled. This is also anticipated in the equation form for Fick's law, given above, since the time factor appears in square root form.

When the solvent is allowed to evaporate from these supported films the reverse process, de-swelling, occurs. The film initially loses solvent quickly and then more and more slowly until the original leached film thickness l is reached again. The compared rates of swelling and de-swelling are illustrated in a typical case in Figure 4-19. It is usually observed that the rate of de-swelling is slower than that of swelling. In this instance the film takes four minutes to reach the half swelling value and more than eight minutes to reach the same level on de-swelling. It is often observed that complex polymeric organic

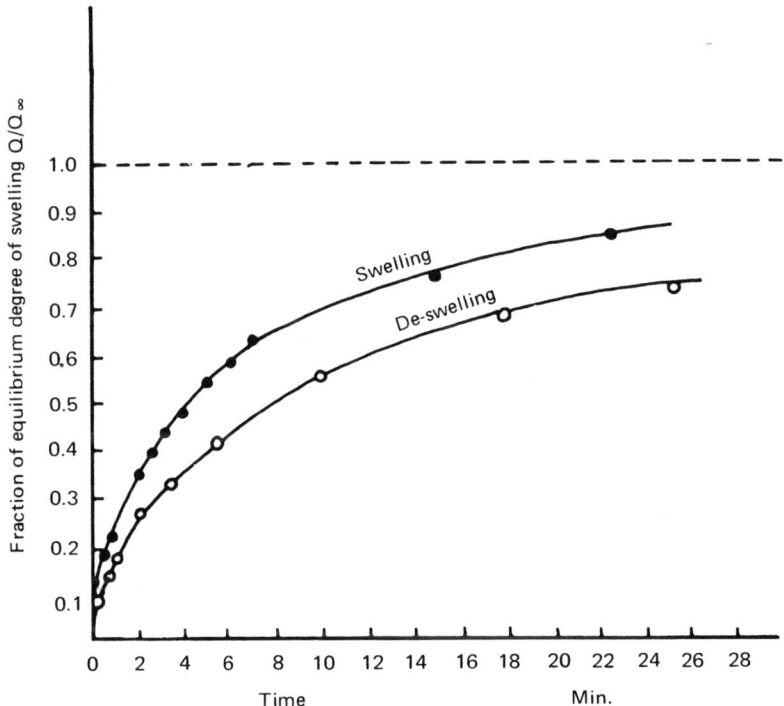

Figure 4-19 Comparison of Rates of Swelling and De-swelling at 22°C. Leached stand oil, aged three weeks at 32°C., film thickness 87μ, solvent— CCl_4 at 22°C.

TABLE 4-15. EFFECT OF FILM THICKNESS ON THE RATE OF SWELLING OF SUPPORTED LEACHED FILMS
(Ref. Figure 4-18)

t min.	Q/Q_∞* $l = 58\mu$**	Q/Q_∞ $l = 70\mu$	Q/Q_∞ $l = 89\mu$
(i) Open pot stand oil, leached, aged for three weeks at 32°C., swelling in carbon tetrachloride at 22°C.			
0.25	0.24	0.16	0.14
0.50	0.33	0.24	0.19
0.75	0.39	0.30	0.23
1.0	0.44	0.34	0.26
2.0	0.61	0.46	0.36
3.0	0.72	0.55	0.43
4.0	0.76	0.62	0.49
6.0	0.80	0.71	0.58
7.0	0.81	0.74	0.63
9.0	0.83	0.77	0.68
12.0	0.84	0.79	0.72
16.0	0.84	0.80	0.74
20.0	0.86	0.81	0.76
60.0	1.00	1.00	1.00
	***$\frac{\Delta V}{V_0} = \frac{75}{58} = 1.29$	$\frac{\Delta V}{V_0} = \frac{104}{70} = 1.48$	$\frac{\Delta V}{V_0} = \frac{126}{89} = 1.42$

t min.	Q/Q_∞ $l = 77\mu$	Q/Q_∞ $l = 138\mu$	Q/Q_∞ $l = 211\mu$
(ii) White lead/stand oil, leached, aged for seventeen weeks at 32°C., swelling in methanol at 32°C.			
1.0	0.27	0.19	0.10
2.0	0.46	0.23	0.15
3.0	0.59	0.30	0.17
4.0	0.68	0.35	0.19
5.0	0.73	0.40	0.21
8.0	0.80	0.50	0.30
10.0	0.82	0.55	0.34
12.0	0.88	0.64	0.38
16.0	0.95	0.68	0.47
20.0	–	–	0.51
25.0	–	0.85	–
30.0	–	–	0.62
40.0	–	0.95	0.72
45.0	–	–	0.78
60.0	1.00	1.00	–
120.0	1.00	1.00	1.00
	$\frac{\Delta V}{V_0} = \frac{11}{77} = 0.14$	$\frac{\Delta V}{V_0} = \frac{20}{138} = 0.15$	$\frac{\Delta V}{V_0} = \frac{26}{211} = 0.12$

Notes: *The ratio Q/Q_∞ represents the fraction of total swelling action that has been completed at time t.

**The thickness before the experiment is given by l which represents the thickness of the leached supported film.

***The equilibrium degree of swelling as defined previously, $\Delta V/V_0$ is the ratio of the increase in volume upon swelling to the unleached volume. Since these films are supported $\Delta V/V_0$ can also be expressed as the ratio of the maximum change of film thickness at swelling equilibrium to the original leached film thickness.

substances, of which linseed oil film is an example, hang on stubbornly to absorbed solvent. The last traces of absorbed solvent evaporate away very slowly over many hours, or even days, depending on its volatility and any specific attraction it may have for the polymer structures. In the laboratory the usual way to remove all of the solvent efficiently is to place the films in a vacuum chamber for an hour or less at about 1 mm. of mercury pressure, and at a temperature close to the normal boiling point of the solvent.

It would be very useful to know how the rate of swelling is affected by the solvent. In general one would expect solvents of higher boiling point and greater viscosity to penetrate films more slowly than those of lower boiling point and greater fluidity. Thus, considering the series of alcohols, n-amyl, n-butyl, n-propyl, ethyl and methyl alcohols, it would be expected that methanol would be the fastest to penetrate (and to leave) the film. The same might be said about other families of solvents, e.g., methyl n-propyl ketone, methyl ethyl ketone, and acetone; or the group diethyl benzene, ethyl benzene, xylene, toluene, and benzene.

The mathematical derivations of the theory of diffusion of solvents into and from polymeric films, such as those of linseed oil, is too detailed to be presented here. Various researchers on cellulose and other plastic films which exhibit swelling behavior have applied Fick's law with some success for the early portions of the diffusion process, e.g., up to the half-absorption (swelling) level corresponding to $Q/Q\infty = 0.5$. The assumption is made here that the diffusion coefficient (measured in units of cm.^{2} per second) is constant up to this level of swelling. In effect, the diffusion coefficient is changing in the initial stages, and what one measures experimentally is an averaged diffusion coefficient.

The resulting relationship between the diffusion coefficient D, film thickness l, and time for half the swelling (or de-swelling process) to take place is found to be:

$$D = \frac{0.2 l_s^2}{t_{0.5}}$$

for supported films where l_s is the supported film thickness in cms. and $t_{0.5}$ is the half-time in seconds. In the case of free films where solvent penetration can occur on both faces, the relationship is:

$$D = \frac{0.05 l_F^2}{t_{0.5}}$$

where l_F is the free film thickness in cms., D throughout is given in units of cm.^{2} per second (cm.^{2} sec.$^{-1}$).

Experimentally it is fairly simple to estimate the film thickness, time for half-swelling; and the diffusion coefficient is calculated. The proviso is that Fick's law applies. That this is so is shown in Figure 4-20, where linear behavior is noted up to and beyond the half-swelling (or de-swelling) stage. The close fit of the data for different film thicknesses (open pot stand oil in carbon tetrachloride) supports quite well the application of Fick's law of diffusion as modified above. It should be emphasized that temperature is an important factor throughout, profoundly affecting the diffusion coefficient (and hence the rate of swelling or de-swelling), and is always specified.

A variety of solvents and film systems was examined experimentally and the results are recorded in Table 4-16. Here both supported and unsupported films are included as well as pigmented and unpigmented ones. In the case of the unsupported films there is a paucity of data for the swelling process, but sufficient data for de-swelling. The measurements were made here gravimetrically, i.e., by following the loss in weight of solvent from the evaporating film as a function of the time. The supported films were measured in the usual way, i.e., on the swelling measurement apparatus by following changes in film thickness with time.

In Table 4-16, Part A, the diffusion coefficients are calculated for supported unpigmented and pigmented films during the swelling stage. Generally the same consistent order is obtained where acetone, benzene, and methanol are the fastest swellers (having the highest values of diffusion coefficient D), and at the bottom of the range are the more viscous, more high-boiling solvents such as n-butanol, n-amyl alcohol, and isooctane. It appears from these data that the pigmented films offer a somewhat greater resistance to solvent penetration in that some of the solvents have slightly lower values of D. The hindrance of the pigment particles, however, would be expected to reduce the diffusion coefficients even further, considering that the white lead films here have an oil content of 16%. Apparently there are sufficient paths of penetration in pigmented films to give solvents more or less direct access to the organic film structure. The comparison of the swelling and de-swelling coefficients is interesting to record; the data for supported but unpigmented films are shown in Part B. The de-swelling process, as has been mentioned earlier, is slower than the swelling, and this is reflected in the lower values of the diffusion coefficient. The de-swelling diffusion coefficients are approximately 40% of the swelling coefficients. Thus, the evaporation of solvents from linseed oil films occurs more slowly than penetration, a phenomenon which has been observed in a number of organic polymer film systems.

In Part C are listed the diffusion coefficients for nineteen solvents as they evaporate away from pigmented white lead/stand oil unsupported films. The de-swelling has been followed here by measuring accurately the changes in weight of film with time. Again, the highly *volatile* solvents and those of low viscosity are the fast evaporators from the films (e.g., methylene chloride,

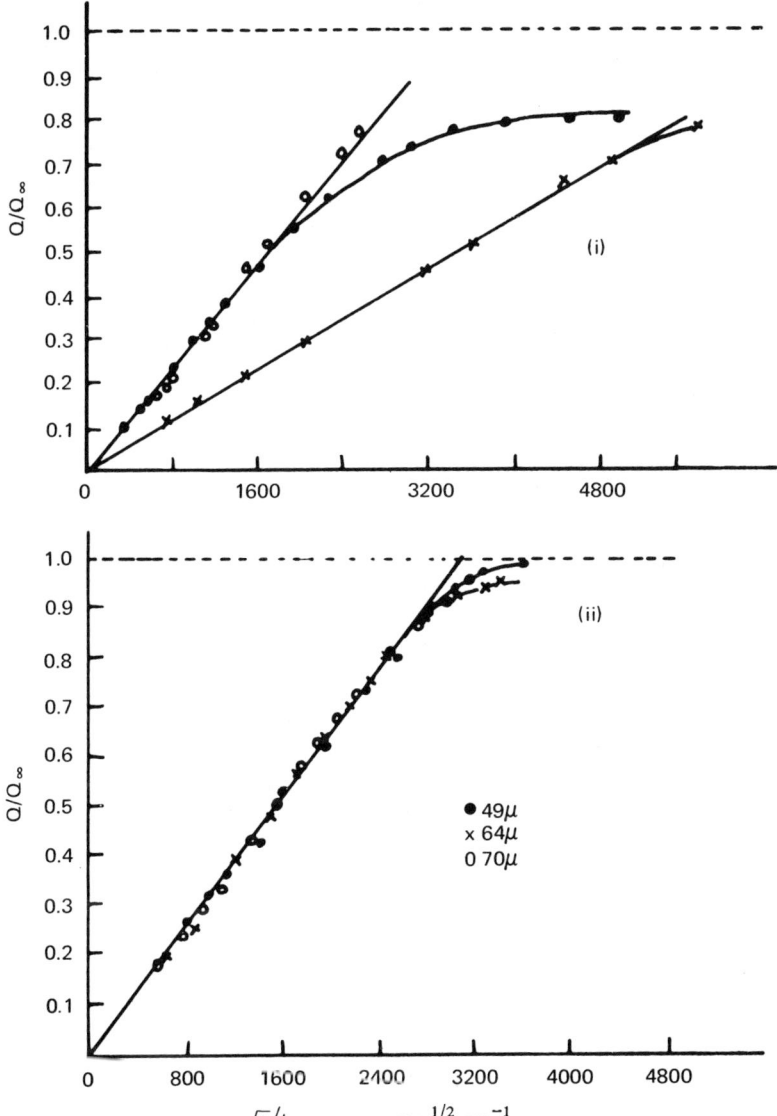

Figure 4-20 The Swelling of Films at 22°C. Tested According to Fick's Law of Diffusion. (i) —o— *white lead/stand oil, eight weeks aging at 32°C., pre-leached, swelling in MeOH.* —●— *stand oil, three weeks aging at 32°C., pre-leached, swelling in CCl$_4$.* —x— *stand oil, forty-eight hours aging at 80°C., pre-leached, swelling in n-amyl alcohol. (ii) Stand oil films, forty-eight hours aging at 85°C., pre-leached, swelling in CCl$_4$.*

TABLE 4-16. DIFFUSION COEFFICIENTS FOR SWELLING AND DE-SWELLING OF SOLVENTS BY LINSEED OIL FILMS AT 22°C.

A. Supported Unpigmented and Pigmented Films—Swelling

FILM TYPE AND AGING	SOLVENT	$l_s \times 10^4$ cm.	$t_{0.5}$ sec.	$D \times 10^8$ cm.^{2} sec.$^{-1}$
(i) Open pot stand oil, aged 72 hrs; 80°C., 3 months normal diffuse light, leached in acetone	Acetone	94	72	25
	Methanol	96	180	10
	Ethanol	96	420	4.4
	n-Propanol	94	720	2.5
	n-Butanol	96	720	2.5
	n-Amyl alcohol	98	1140	1.6
(ii) Alkali-refined linseed oil 0.5 Pb (as naphthenate) aged 72 hrs; 80°C., 3 months normal diffuse light; leached in acetone	Acetone	76	45	26
	Methanol	77	84	14
	Ethanol	76	270	4.3
	n-Propanol	75	480	2.4
	n-Butanol	76	720	1.6
	n-Amyl alcohol	77	900	0.8
(iii) White lead/stand oil, aged 12 months 32°C., normal diffuse light; leached in acetone; 84% pigment, 16% oil content	Acetone	84	90	16
	Benzene	74	108	10
	Methanol	110	240	10
	Ethanol	168	1500	4.2
	n-Propanol	170	1560	3.4
	n-Hexane	103	600	3.0
	n-Butanol	172	2400	2.5
	n-Heptane	103	960	2.0
	Isopropanol	103	1800	1.2
	n-Amyl alcohol	72	1800	0.9
	Isobutanol	74	2100	0.6
	Isoamyl alcohol	132	5400	0.6
	Isooctane	213	15600	0.6

B. Comparison Swelling and De-swelling Diffusion Coefficients—Supported Films

FILM TYPE AND AGING	SOLVENT	SWELLING $D \times 10^8$ cm.^{2} sec.$^{-1}$	DE-SWELLING $D \times 10^8$ cm.^{2} sec.$^{-1}$
Open pot stand oil, 48 hrs; aging at 85°C., then 1 yr. at 32°C.; leached in acetone	Acetone	36	14
	Ethylene dichloride	33	—
	Benzene	30	7
	Methanol	12	4.6
	Carbon tetrachloride	9	3.8
	Ethanol	4.3	2.3
	n-Propanol	2.0	1.1

TABLE 4-16.—Continued

C. Unsupported Pigmented Films—De-swelling Diffusion Coefficients*

FILM TYPE AND AGING: white lead/stand oil, 84% pigment, 16% oil content; leached in methanol and acetone; aged 14 years 20–30°C., normal diffuse light, thickness 170μ

SOLVENT	DE-SWELLING $D \times 10^8$ cm.^{2} sec.$^{-1}$	SOLVENT	DE-SWELLING $D \times 10^8$ cm.^{2} sec.$^{-1}$
Methylene chloride	48	Ethyl benzene	4.4
Tetrahydrofuran	32	Isopropanol	4.2
Chloroform	32	Ethanol	4.0
Ethylene dichloride	32	Cellosolve	3.5
Benzene	24	n-Propanol	3.0
Acetone	18	n-Butanol	2.6
Toluene	18	n-Amyl alcohol	1.8
Xylene (o, m, p)	8	Isooctane	1.4
Methanol	5.5	Diethyl benzene	0.9
Methyl cellosolve	5.5		

*The data of Part C, for unsupported, or free, films, were obtained by weight changes measured on a highly sensitive balance.

NOTES: The diffusion coefficients are listed in descending order, that is with the fast diffusing solvents first, followed by the slower ones. All diffusion measurements calculated at 22°C. from the observed film thickness, and the time for half-swelling (or de-swelling)-$t_{0.5}$.

tetrahydrofuran, chloroform, ethylene dichloride, acetone, benzene), and those of lower volatility and greater viscosity (e.g., n-amyl alcohol, isooctane, and diethyl benzene), are the slow evaporators. Comparison of the values of D for the de-swelling of free films with those for de-swelling of supported films tends to show that free films de-swell faster than supported films, all other conditions being equal. It may be that the film structure for supported films, being under some restraint, has something to do with these results. It also appears from other experiments, not detailed here, that the differential between swelling and de-swelling diffusion for free unsupported films is not nearly so great as that for supported films (cf. Part B of Table 4-16).

The data of Part C, on comparison with published evaporation rates for solvents,[48] show some broad degree of conformity in that the solvents with high volatility are those that diffuse fastest into and from the linseed oil films. There are exceptions, however, which suggest other factors more related to specific attractions that certain solvents have for the linseed polymeric structure.

An attempt to relate diffusion coefficients to such factors as molecular volume and viscosity is demonstrated in Figure 4-21. The plots for white

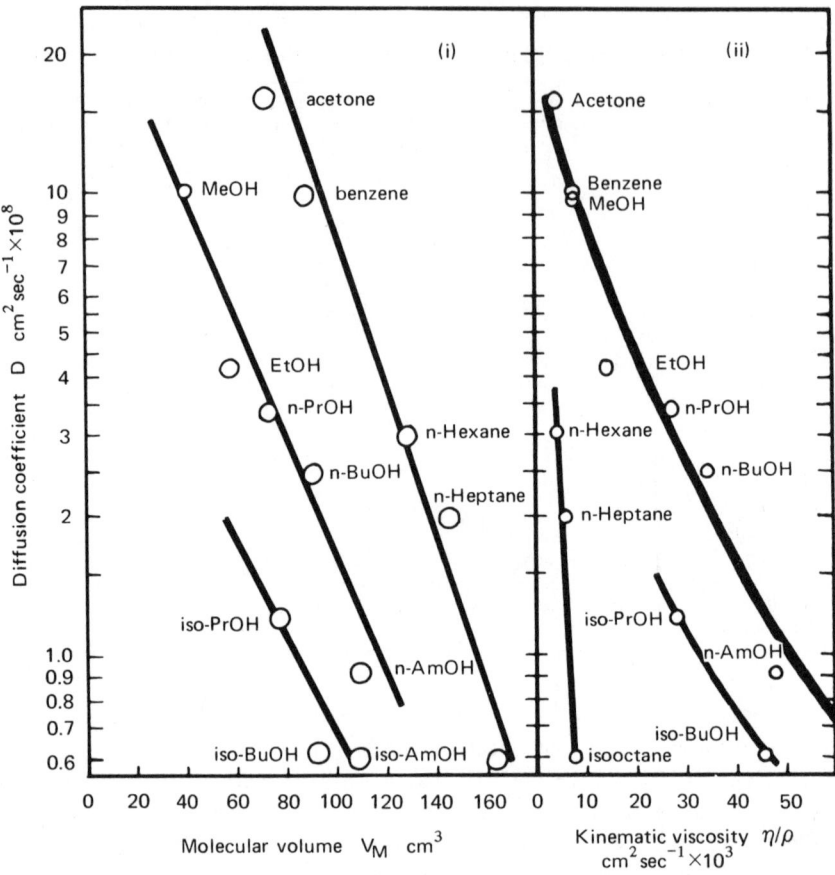

Figure 4-21 The Plot of Diffusion Coefficients: against (i) the molecular volume, (ii) the kinematic viscosity of the solvent. The films are white lead/stand oil, aged twelve months 32°C., swelling at 22°C.

lead/stand oil films at 22°C. show that the solvents tend to fall into consistent classes. In the plot of molecular volume V_M (the molecular weight divided by the density of the solvent) it is seen that the straight chain alcohols (e.g., methanol, ethanol, n-propanol, etc.) fall on one line, that acetone and the aromatic and aliphatic hydrocarbons (e.g., benzene, n-hexane, n-heptane) fall on another line, while the alcohols with branched chains (isopropanol, isobutanol, and isoamyl alcohol) form another consistent group. From this kind of analysis it seems that for equivalent molecular volume, solvents such as acetone and the hydrocarbons diffuse faster than the alcohols, and these in turn diffuse faster than the branched chain alcohols.

In the second portion of Figure 4-21 consideration is given to the viscosity of the solvent, in particular, the kinematic viscosity, which is the viscosity divided by solvent density. This time acetone and benzene fall into the same group as the normal alcohols, and the hydrocarbons are distinctly separate. From this it is seen that while the hydrocarbons have low viscosity, their degree of penetration or diffusion is generally lower than the oxygen containing solvents.

The graphical presentation in Figure 4-21 demonstrates the operation of physical penetration of solvent into the films with class differences. These class differences indicate that in addition to physical differences there are specific attractions that certain solvent types have for the linseed oil structure. The fact that the leached films can always be readily de-swelled suggests that any specific interaction between the solvent and the film structure is weak and can be undone.

In this discussion so far the temperature itself was not considered a variable. The effect of increasing the temperature is to cause a dramatic increase in the diffusion coefficient, and vice versa. This is entirely to be expected in such systems.[49] An increase of 20° in temperature increases the diffusion coefficient by about 80% or more. Even fast-swelling solvents such as methylene chloride can be reduced in their degree of penetration by lowering the temperature, say, from 22° to 10°C., effectively reducing the diffusion coefficient by about 50%.

The degree of swelling at equilibrium, it should be emphasized, does not appear to be influenced to any appreciable degree by change in temperature.

GENERAL CONCLUSIONS

As discussed in the Introduction, the study in such detail of the leaching, swelling, and diffusion properties of solvents in linseed oil films was designed to better our understanding of such systems. In particular, these studies were carried out to apply the results, if possible, to the cleaning of works of art. The information obtained from the study of carefully controlled experiments on pigmented and unpigmented films, some of considerable age, leads one to certain definite conclusions.

One of the first conclusions is that the cleaning of a virgin oil paint surface will undoubtedly lead to irreversible changes. Not only will there be leaching from the surface of the paint but in the process of rubbing with a cotton-wool swab some losses of medium will occur. This may serve to explain the chalky condition of some cleaned pictures where there is a greater diffuse scatter of light from pigment particles freshly exposed to the air. While the leaching action cannot be remedied, it is possible to minimize the mechanical action.

Often it appears necessary to employ solvents in the cleaning of pictures which have potentially great penetrating power and high swelling action on the underlying paint. Recourse to solvents such as acetone is frequently necessary in the removal of certain types of varnish coatings. (Ruhemann discusses this at some length.) In some cases methylene chloride is used, which is related to chloroform in its high swelling action on dried linseed oil. The use of such solvents presents great dangers unless the nature of diffusion and swelling is clearly understood. Thus, one may use acetone with relative safety if the time of contact with the varnish surface is minimized, small quantities are used, and adequate time for evaporation allowed. Repeated solvent application in one area will build up dangerously high concentrations of solvent within the original paint layers, to such an extent that even slight mechanical action may prove to be very damaging.

Ideally, for safe removal of varnishes from an oil painting in the future, it would be well to employ a varnish that is removable in solvents having low swelling and diffusing action on oil paint, solvents such as the paraffinic hydrocarbons which have been shown to have low swelling action and low diffusion rate (cf. Figures 4-15 and 4-21.) So long as materials such as mastic and dammar are used, the aged films of which usually require oxygen containing solvents (e.g., ethyl alcohol, acetone, diacetone alcohol) for their removal, some danger in picture cleaning would be experienced. These solvents, of course, tend to have strong action on linseed oil films as regards leaching and swelling.

In picture cleaning one should balance the two factors of swelling power of solvent and rate of diffusion. The swelling power can be gauged beforehand with reasonable accuracy by a consideration of molecular volumes (Figure 4-14) or, more effectively, by applying solubility parameters (Figure 4-15). The rate of diffusion can be roughly predicted from the viscosity or molecular volume of the solvent. In some particular cases in picture cleaning, it might prove feasible to employ a fast-penetrating solvent of low swelling action (e.g., methyl alcohol); in other cases, to employ one of slow penetration and high swelling action (e.g., normal butyl alcohol).

Different modes of solvent action on films are shown in Figure 4-22. In this figure are three equivalent films of dried linseed oil. Each is exposed to a different solvent for a period of ten seconds. The film on the right has been softened to an appreciable extent, since a fast-diffusing and high-swelling solvent, such as benzene or ethylene dichloride, has been used. The film on the left has been treated with isooctane, or isopropyl alcohol, and the softening is low. If both films are mechanically rubbed, the wearing of the surface would

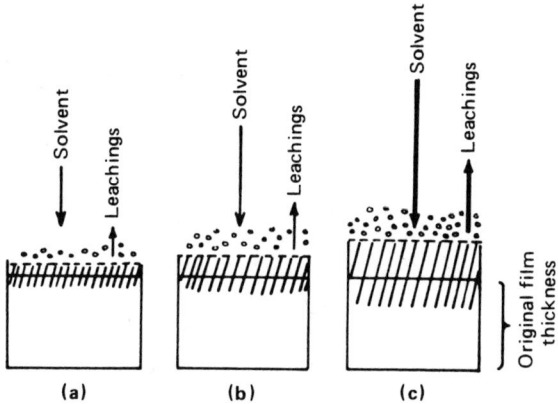

Figure 4-22 Different Modes of Solvent Action. (a)–Slow diffusion, low swelling action, low leaching rate, e.g., isooctane. (b)–Moderate rate of diffusion, moderate swelling action, appreciable leaching rate, e.g., methyl alcohol. (c)–Fast rate of diffusion, great swelling action, high leaching rate, e.g., ethylene dichloride, benzene.

be less noticeable in the second case. The film in the center refers to an intermediate condition. Thus the damage to be expected on an oil painting surface during cleaning depends on the depth of penetration of solvent and on the swelling level. The leaching of soluble components would nevertheless occur in every case; the degree of leaching would depend on the time of contact of solvent, the type of solvent used, film thickness, and the nature of the film itself (e.g., pigment and composition). The danger of leaching lies of course in its irreversibility.

One cannot extrapolate these results to films of other media, e.g., tempera or oleoresinous compositions. Research into the action of solvents on such media is of course very necessary, and these results, together with the known results on linseed oil, would form a very useful corpus of knowledge which would give the conservator much more confidence in picture cleaning.

There are, of course, certain complicating factors. Very often one is confronted with paintings where the medium is not known, or it may vary within the paint film, or from one area to another. In order to remove varnishes with greater security in these instances, a chemical knowledge of the composition of the medium is necessary. The application of gas chromatography described here in some detail could prove most useful. As more results accumulate through identification studies of media of old works of art by research efforts such as these, the practice of picture cleaning cannot help but benefit. New tests will have to be devised to give the picture cleaner more objective guidance regarding film swelling and leaching action. A decision on the course to be taken in picture cleaning would be based on the method producing the minimum of swelling and leaching action. In some cases it will be considered safer not to attempt varnish removal at all.

As regards further directions that would prove fruitful in this line of research, it would be useful to elucidate the nature of those components in the leachings which do not yield esters of monocarboxylic (e.g., palmitic, stearic) and dicarboxylic acids (e.g., suberic, azelaic). It would be useful to carry out a series of experiments on control paintings duplicating the various methods of cleaning action and analyzing the swabs for leached products. In this respect the effect of pigmentation would be given special attention. Finally, sampling procedures should be devised which allow for the accumulation of paint fragments and oil grounds so that more statistical evidence can be accumulated on that as yet unexplored territory of the artists' materials, the medium.

SUMMARY

Varnish films yellow and deteriorate with age thus requiring removal by cleaning with solvents. The possibility of solvent action upon underlying paint films during cleaning is the subject of this study. Previous research in this field is very sparse; although accounts in the past century warn against the action of certain solvents during cleaning of paintings.

Linseed oil films dry by a polymerization process in which the unsaturated triglyceride molecules combine with the aid of oxygen. At the same time, deterioration occurs by the splitting or scission at the sites of oxygen linkages in the film structure giving rise to a variety of decomposition products such as, dicarboxylic acids (e.g., pimelic or azelaic acids), ketones, aldehydes, alcohols, carbon dioxide and water vapor. These materials together with unpolymerized triglycerides (composed essentially of stearic and palmitic acid units) form a kind of interstitial plasticizer in the dried linseed oil film. Where pigment is present some modification of the drying process may take place; certain pigments function as catalysts, and others as retarders. The penetration rate of solvent will of course be slower in pigmented films than unpigmented ones. Also the photochemical action of light upon the medium locked in by the pigment may depend on the light-absorbing properties of the pigment itself.

When solvent is brought into contact with a dried linseed oil film the solvent rapidly diffuses into the film causing it to swell. This results in the almost instantaneous loss of soluble or leachable components. After a period of time (usually minutes), the leaching process is almost complete with the film still remaining in a swollen and softened condition. When the solvent is allowed to evaporate, it is observed that the film has lost weight, decreased its volume, has become brittle, and increased in density. This is attributed to the leaching of soluble components from the film which are identified with oxidation products, saturated triglycerides, and low molecular weight polymers of the triglycerides. Repeated solvent contact leads mainly to swelling action. Further leaching depends on the history of solvent action on the film and aging processes. In tests carried out on varnished paint films the swelling and leaching action were not measurably reduced. According to film age and type maximum leaching losses are of the order of 20-50%, about half of these values being reached in 100 sec. for a solvent such as acetone. In pigmented films where the drying process is very slow or where film oxidation of the structure predominates (such as films containing iron oxide or titanium dioxide), solubilities of the order of 80% are found. This results in disruption of the paint film and loss of pigment particles. It might be expected that increasing film age should dramatically reduce solubility losses upon solvent contact; however, this has not been found. Paint films up to 300-years old exhibit leaching or solubility behavior although swelling may be at lower levels than in younger films. Inert solvents such as n-hexane leach less material than such oxygen-containing solvents as acetone, ether, methanol and cellosolve. This applies also to high-swelling, polar-chlorinated solvents such as methylene chloride, ethylene dichloride, or chloroform. However, above a certain threshold degree of swelling the amount of leaching is not increased significantly suggesting that the linseed oil film structure is composed of a three-dimensional network entrapping lower molecular weight materials described earlier.

The nature of the leached components has been studied by infra-red spectroscopy and by gas chromatography, which show that they are unreacted or low molecular weight triglycerides containing palmitic or stearic acid, dicarboxylic acids resulting from oxidation film breakdown processes, along with hydroperoxides, aldehydes, ketones, and alcohols. A semi-quantitative study of the relative amounts of these components appears to confirm the random distribution theory for the various triglycerides in the original oil. The unchanged palmitic and stearic acid triglycerides, along with such dicarboxylic acids as azelaic or pimelic acids, are readily estimated on a quantitative basis by esterification to methyl esters of the leachings. Time studies of leaching action show that the lowest molecular weight species such as alcohols, ketones, aldehydes are removed first, followed by saturated triglycerides, and, finally, by dicarboxylic acid components. In general it has been possible at least by gas chromatography to analyze the esters of the leachings of oil films dating from the seventeenth century.

The swelling action of unpigmented and pigmented films was studied in some detail by special instrumental techniques. A suitable interpretation of the results can be made by applying the theory of solubility parameters. These numbers range from 5 for fluorocarbons to 23.2 for water and are linked to certain basic solubility properties of solvents in general. Polymeric substances also have solubility parameters and where they match those of the solvent, maximum swelling or solubility can occur. White lead/stand oil films aged up to fourteen years have similarly been studied and the solubility parameter of the dried oil is found to be between 9.3-9.5 units. A total of fifty-six solvents was tested for swelling action. Strong swelling solvents are ethylene chloride, trichloroethylene, methylene chloride, chloroform, tetrahydrofuran, cyclohexanone, having solubility parameters in the range of 9.3-9.9 units. Low swelling solvents are those having parameters well below 9.3 or well above. In this category are the petroleum solvents and most of the alcohols. It is emphasized that regardless of low swelling action, considerable leaching may still occur, such as with methanol. Solvents with intermediate swelling action are acetone, cellosolve, benzene and carbon tetrachloride. The effect of mixing of solvents can also be predicted. Thus solvents chosen from either side of the "maximum" of the solubility parameter/solvent curve can enhance one another when combined to give stronger swelling power. This has been observed, for example, when mixing benzene and methanol.

The rate of swelling, or the diffusion of solvents into and from linseed oil films, has been studied in some detail. This process was found to occur according to Fick's law. The thickness of film has a profound effect on the time for a given amount of swelling to take place, e.g., doubling the thickness increases the time by a factor of four, and vice versa. Supported films take longer to reach swelling equilibrium than do unsupported films. The evaporation of solvent from films generally proceeds at a slower rate than the penetration process. The last traces of solvent in the film are very slow to leave, stubbornly retained, and conveniently removed by vacuum evaporation. The calculated diffusion coefficients are given for various solvents (measured as cm.^{2} per second) with reference to a fixed temperature. Thus for supported white lead/stand oil films of thickness 80-100μ the diffusion coefficients at 22°C. are 16×10^{-8} cm.^{2}/sec. for acetone and 0.6×10^{-8} cm.^{2}/sec. for isooctane. In terms of time for half the swelling process to be reached, acetone—90 sec., isooctane—2100 sec. In a series of evaporation studies the order of evaporation was: methylene chloride, tetrahydrofuran, chloroform, ethylene dichloride, benzene, acetone, toluene, xylene (o,m, and p), methanol, methyl cellosolve, ethyl benzene, isopropanol, ethanol, cellosolve, n-propanol, n-butanol, n-amyl alcohol, isooctane, and diethylbenzene. In general, the more volatile the solvent the faster is its penetration (and evaporation) into (or from) linseed oil films. The effect of temperature on these factors is very pronounced. An increase of 20°C. in temperature increases the diffusion coefficients by 80% or more, and vice versa. The amounts of maximum swelling are not significantly affected by temperature as are the rates of attainment of swelling to the maximum or equilibrium values.

The conclusions drawn from the various studies are important and far-reaching. The cleaning of a virgin oil paint surface (i.e., one never previously in contact with solvent) will lead to leaching action which is irreversible. This may serve to explain the "chalky" condition of some cleaned pictures. While leaching can not be eliminated entirely, mechanical action upon softened films resulting from swelling can be minimized. Acetone may be used relatively safely if the contact with the surface is minimized, small quantities are used, and adequate time for evaporation is allowed. Repeated solvent application in one area will build up dangerously high concentrations of solvent within the original paint layers so that mechanical action on cleaning may be very damaging. The twin factors of swelling and rate of diffusion should be balanced. Both can be determined from the data given for the various common solvents. Different modes of action are also given. These include employing a fast diffuser and high sweller such as benzene or ethylene dichloride, a slow diffuser and low sweller such as isooctane or isopropyl alcohol, or in an intermediate situation, a medium fast diffuser and medium sweller such as methanol. Regardless of mechanical damage by rubbing of such swollen films, the leaching of soluble components will take place. The danger of this is its irreversibility.

Ideally for future safe removal of varnishes from oil paintings it would be necessary to employ varnishes which are removable in solvents having low swelling and diffusing action to the paint film. As long as varnishes such as mastic and dammar are used, the aged films of which require application of oxygen-containing solvents, e.g., ethyl alcohol, acetone, etc., some danger will be experienced through leaching and swelling. In some cases varnish removal should not be attempted at all.

Broader directions in the research indicated are to elucidate further the nature of the leached components, to develop simple test methods for detection of these components on solvent swabs, and to develop sampling and analytical procedures for increasing our knowledge of "oil" paintings.

REFERENCES AND NOTES

1. Watin, *L'Art du Peintre, Doreur, Vernisseur*, 4th ed. (Paris, 1793).
2. H. Déon, *De la Conservation et de la Restauration des Tableaux* (Paris, 1851).
3. *Report from the Select Committee on the National Gallery—Together with the Proceedings of the Committee, Minutes of Evidence, House of Commons* (London, 1853).
4. N. Brommelle, "Material for a History of Conservation," *Studies in Con-*

servation 2 (1956): 176-88. The author refers largely to the report of Reference 3 and describes various hazardous and unorthodox methods of cleaning.
5. R. H. Marijnissen, *Degradation, Conservation, et Restauration de l'Oeuvre d'Art*, 2 vols. (Bruxelles: Arcade, 1967).
6. H. Ruhemann, *The Cleaning of Pictures – Problems and Potentialities* (London: Faber & Faber, 1968).
7. Reference to Faraday's experiments and his comments on cleaning are to be found in the report of Reference 3, in particular, Minutes no. 5551-55, 5472-5508, 5556-59, and especially 5524–"Q: Do you consider that solvents, very much diluted, and in careful hands, may safely be applied to the surfaces of pictures with a view to the removal of decayed varnishes and impurities? A: If a solvent be employed, and there is dirt upon the picture, it will be removed by the solvent power, but if sufficient solvent power be retained to remove the varnish, that which is beneath, which consists of part oil and part varnish, will also be partly removed. If there were little varnish and much oil, there would be very little danger in it; but one is always in the condition of passing gradually and unawares from danger to security, and from security to danger, inasmuch as whatever will remove varnish will remove in some degree the under parts of such a picture."
8. A. P. Laurie, "Restrainers and Solvents Used in Cleaning Old Varnish from Pictures, *Technical Studies in the Field of the Fine Arts* 4, No. 1 (1935): 34-35.
9. M. Von Pettenkofer, *Über Ölfarbe und Conservierung der Gemälde-Gallerien durch das Regenerations-Verfahren* (Braunschweig: Vieweg, 1872).
10. U. Forni, *Manuale del Pittore Restauratore* (Florence, 1866).
11. See Chapter 8 of this book.
12. G. L. Stout, "A Preliminary Test of Varnish Solubility," *Technical Studies in the Field of the Fine Arts* 4, No. 3 (1936): 146-61.
13. A. P. Laurie, "The Selective Sorption of Organic Liquids by Solid Films of Raw Linseed Oil and Stand Oil," *Transactions of the Faraday Society* 33 (1937): 293-99.
14. A. P. Laurie, "Restrainers and Solvents Used in Cleaning Old Varnish from Pictures," *Studies in the Field of the Fine Arts* 4, No. 1 (1933): 34-35; also A. P. Laurie, "Le Dévernissage des Tableaux Anciens et la Suppression des Repeints," *Mouseion* 25-26 (1934): 216-19.
15. H. Ruhemann, *Manuel de la Conservation et de la Restauration des Tableaux* (Paris: L'Office International des Musées, 1939), p. 128. This Manual first appeared in 1938 as vols. 41-42 of *Mouseion*, official organ of L'Office International des Musées of the old League of Nations.

16. S. Rees Jones, An unpublished memorandum suggesting that diffusion and swelling are basic factors to consider in picture cleaning (London: Courtauld Institute of Art, 1949).
17. *An Exhibition of Cleaned Pictures 1936-1947*, rev. ed. (London: National Gallery, 1947). This is the catalog of the exhibition.
18. The Weaver Report, "The Care of Paintings," *Museum* (Paris: UNESCO, 1949-50): 9-31; being a reprint of the original report, J. R. H. Weaver, G. L. Stout, and P. Coremans, "Report of a Committee of Confidential Inquiry into the Cleaning and the Care of Pictures in the National Gallery," mimeographed (London, 1947), iii: 54 pp.
19. N. Stolow, "Some Investigations of the Actions of Solvents on Drying Oil Films" (Ph.D. diss., University of London, 1955), 261 pp.; reprinted in two parts by N. Stolow in *J. Oil and Colour Chemists' Assoc.* 40 (1957): Part I, 337-402; Part II, 438-99.
20. I. Graham, "The Effect of Solvents on Linoxyn Films," *J. Oil and Colour Chemists' Assoc.* 36 (1953): 500-506.
21. N. Stolow, "Application of Science to Cleaning Methods: Solvent Action Studies on Pigmented and Unpigmented Linseed Oil Films," in *Recent Advances in Conservation—Contributions to the IIC Rome Conference, 1961* (London: Butterworths, 1963), pp. 84-88.
22. N. Stolow, "The Application of Gas Chromatography in the Investigation of Works of Art," in *Application of Science in the Examination of Works of Art—Proceedings of the Seminar Held at the Museum of Fine Arts, Boston, 1965* (Boston: Museum of Fine Arts, 1967), pp. 172-83.
23. H. J. Dutton and J. A. Cannon, "Glyceride Structure of Vegetable Oils by Countercurrent Distribution. I—Linseed Oil," *J. American Oil Chemists Society* 33 (1956): 46-50.
24. H. J. Dutton and C. R. Schofield, "Recent Developments in the Glyceride Structure of Vegetable Oils," *Progress in the Chemistry of Fats and Other Lipids* 6 (1963): 313-39.
25. F. D. Gunstone and R. B. Padley, "Glyceride Studies. Part III. The Component Glycerides of Five Seed Oils Containing Linolenic Acids," *J. American Oil Chemists Society* 42 (1965): 957.
26. A. G. Vereschagin and G. V. Novitskaya, "The Triglyceride Composition of Linseed Oil," *J. American Oil Chemists Society* 42 (1965): 970-74.
27. The Iodine Value (I.V.) of an oil or fat is the percentage by weight of iodine absorbed under prescribed test conditions. The most widely used is the one based on the method of Wijs, agreed upon by the International Commission for the Study of Fats—and detailed in specified test methods of A.S.T.M. and B.S.I. (e.g., *B.S.S.* 684, London, 1950, p. 82).
28. G. H. Hutchinson, "Some Recent Advances in the Chemistry and Tech-

nology of Drying Oils," *J. Oil and Colour Chemists' Assoc.* 41 (1958): 474-92.
29. N. Stolow, "A Modified Apparatus for Measuring the Swelling of Polymer Films in Solvents," *J. Scientific Instruments* 31 (1954): 416-20, in which the method of measuring the swelling of supported films is described in detail. The unsupported films were measured with a travelling microscope accurate to within ±0.05 mm or better. The horizontal measurements were made on the film after gently flattening with a glass plate. This was found to offer no restraint to the swelling action during the short interval of measurement. The swelling film thickness of unsupported films was measured by means of a dial gauge fitted with a 1/2" diameter flat probe sensitive to displacements of $\pm 1\mu$ or better.
30. L. Masschelein-Kleiner, J. Heylen, and Tricot-Marckx, "Contribution á L'Analyse des Liants, Adhésifs et Vernis Anciens," *Studies in Conservation* 13, No. 3 (1968): 105-21.
31. The density of the leached components may be obtained from the relationship (cf. Ref. 19):

$$\frac{D}{d_L} + \frac{1-S}{d_F} = \frac{1}{d_i}$$

where d_L, d_F, and d_i refer respectively to the densities of the leached components, the film after leaching, and the film before leaching; S refers to the observed leaching loss in weight of 1 g. of original film. Experimentally it is always observed that $d_F > d_i$ and hence d_L is always lower than d_i.
32. The instrument used for this purpose was a Perkin-Elmer Infrared Spectrophotometer Infracord Model. The silver chloride plates were positioned in the path of the infra-red beam, i.e., in the normal way for measurements. In other cases the attenuated total reflectance technique proved to be useful.
33. S. O. Crecelius, R. E. Kagarise, and A. L. Alexander, "Drying Oil Oxidation Mechanism, Film Formation, and Degradation," *Industrial and Engineering Chemistry* 47 (1955): 1643-49.
34. C. R. Bragdon, ed., *Film Formation, Film Properties and Film Deterioration – A study by the Research Committee of the Federation of Paint and Varnish Production Clubs* (New York: Interscience, 1958).
35. R. S. Yamasaki, "Chemical Kinetics of Photo-oxidative Degradation of Dried Trilinolein Film," *Paint Technology* 39 (1967): 134-43.
36. P. L. Jones, "The Leaching of Linseed Oil Films in Iso-Propyl Alcohol," *Studies in Conservation* 10 (1965): 119-29.
37. J. S. Mills, "The Gas Chromatographic Examination of Paint Media.

Part I. Fatty Acid Composition and Identification of Dried Oil Films," *Studies in Conservation* 11, No. 2 (1966): 92-107.
38. B. F. Daubert, "The Composition of Fats," *J. American Oil Chemists Society* 25 (1948): 425. The method of calculating the compositions of triglycerides of different combinations is worked out here according to a random distribution pattern. Thus if the three triglycerides are possible AAA, AAB, ABC, then

$$\%AAA = (A)^3 \times 100;$$
$$\%AAB = 3(A)^2 \times (B) \times 100;$$

and
$$\%ABC = 6(A)(B)(C) \times 100.$$

From the known composition of the five components of linseed oil, i.e., palmitic, stearic, oleic, linoleic and linolenic acid components, it is possible in this manner to calculate the percentage of tripalmitin, tristearin, and the various others of the twenty-one other principal triglycerides.
39. J. Hildebrand and R. Scott, *The Solubility of Nonelectrolytes* (New York: Reinhold, 1949) pp. 129, 361.
40. H. Burrell, "Solubility Parameters," *Interchemical Review* 14 (1955): No. 1, pp. 3-16; No. 2, pp. 31-46.
41. H. Burrell, "The Challenge of the Solubility Parameter Concept," *J. Paint Technology* 40 (1968): 197-208.
42. J. L. Gardon, "The Influence of Polarity upon the Solubility Parameter Concept," *J. Paint Technology* 38 (1966): 43-57. In this paper as in that of Burrell's (Ref. 41) there is considerable discussion of other factors which tend to modify the solubility parameter concept—such as hydrogen bonding and polarity of the solvent.
43. R. M. Barrer, *Diffusion In and Through Solids* (Cambridge: University Press, 1941).
44. J. Crank, *The Mathematics of Diffusion* (London: Oxford University Press, 1956).
45. S. Prager and F. A. Long, "Diffusion Coefficients from Sorption Data," *J. American Chemical Society* 73 (1951): 4072-75.
46. J. Crank (see Ref. 44)- detailed discussion on the diffusion characteristics of swelling plates (e.g., films), pp. 241-42.
47. P. Drechsel, J. L. Hoard and F. A. Long, "Diffusion of Acetone into Cellulose Nitrate Films and Study of the Accompanying Orientation," *J. Polymer Science* 10 (1953): 242-52.
48. L. D. Wilson, "Evaporation Rates of Solvents and an Improved Method for Their Determination," *Oil and Chemical Review* 118, No. 24 (1955): 6-8.

49. According to Barrer (Ref. 43) and others an Arrhenius-type equation for activated diffusion can be applied:

$$D = A \exp(-E/RT)$$

where T is the absolute temperature in °K., E is the activation energy in calories per mole, and R is the gas constant of value 2 cal./mole °K. The activation energies found for linseed oil films were of the order of 5 kcal./mole. Activation energies of the order 5 to 8 kcal./mole are frequently encountered with polymers which exhibit a positive temperature coefficient of the rate of sorption.

Part III

RESINS AND THE PROPERTIES OF VARNISHES

Robert L. Feller

5
THE NON-VOLATILE COMPONENT

The film-forming materials used in formulating solvent-type varnishes usually exhibit the minimum characteristics of crystallinity. They are *amorphous*, i.e., they tend to exhibit conchoidal fracture, optical clarity in mass, no sharp melting point, and virtually no x-ray diffraction pattern (the latter is similar to the result obtained with liquids). Such substances are effectively "solids" by virtue of their flow resistance. They may have characteristics of fluids of such high viscosity that they do not flow under ordinary conditions, or they may exhibit plastic flow (see below). Some of their properties, such as brittle point and second-order-transition temperature, have been studied profitably from the viewpoint of their being highly viscous liquids or glasses; their other properties, such as the effect of solvents and plasticizers on film strength, have been studied successfully from the view of their being gels.[1,2]

Plastic flow requires a definite shearing stress (yield value) before flow can be induced. This is the origin of the term "plastic," a substance capable of being deformed and remaining in the deformed state when the force of deformation is removed; the use of this word implies such behavior. For want of a better word to designate a vast array of man-made materials, and to imply their characteristics only in a very general manner, the word "resin" is widely used; its meaning is no longer restricted to the secretions of plants. The word "polymer," on the other hand, refers specifically to the chemical structure of the resin or plastic, a compound in which a large number (poly) of identical or similar atoms or groups of atoms (representing the *monomer*) are united by primary chemical bonds. Perhaps it might be said that, the less that is known about the chemical structure of a resin-like material, the more likely it will be called a resin. Cellulose, vegetable polysaccharides, lignin, and many proteins are properly called natural polymers, while other substances, natural and synthetic, are still designated simply as resins.[3] Thus, although the three terms are used interchangeably, the term *resin* is relatively non-committal, implying amorphous behavior and no precise designation of the chemical structure; *plastic* implies a type of physical behavior and hence is slightly more restricted in meaning; *polymer* refers to a specific type of chemical structure.

The present chapter shall discuss organic substances, compounds containing carbon. However, inorganic polymers are known, based on sulfur, selenium, silicon, germanium, and other elements. The technology of handling these is not developed as extensively as that of organic polymers, and except for perhaps silicone polymers, they do not as yet appear to offer as great a utility in the protection of traditional paintings as do the polymers containing carbon, hydrogen, and oxygen. Nitrogen-containing polymers have a role to play in conservation, particularly the natural ones in glue, gelatin, and casein. They will not be discussed on this occasion, however, because vinyl acetate and acrylic polymers are highly stable thermoplastic polymers that show promise in conservation and we wish to concentrate primarily on these. These can serve as the standards against which other polymers used in conservation may be compared.

For the sake of convenience, natural resins will be considered separately from synthetic ones.

NATURAL RESINS

Natural resins are often classified into two types: (a) hard, fossil resins which must be thermally processed in order to dissolve them; and (b) those which may be dissolved without heat. The latter are used in solvent-type varnishes. Gettens and Stout, in their book on *Painting Materials*, provide an excellent general outline of the characteristics of these materials and what little is known of their long history in the fine arts.

The composition of natural resins tends to be variable and is seldom completely known. Improved control of their composition is obtained sometimes by isolating certain fractions, as in the precipitation of "dammar wax," or by altering their chemical constitution, as in hydrogenation.[4]

Use of mastic resin in various crafts has been traced to at least the first millennium, B.C. Its use in solvent-type picture varnish is perhaps limited to a few hundred years; dammar picture varnish was introduced by Lucanus in 1829.[5] Prior to this, oil-resin varnishes were prevalent.[6]

Dammar is the general name given to a group of natural resins originating from the *Dipterocarpaceae* family of trees chiefly in Malaya, Indonesia, and the East Indies. Mastic resin is derived from the tree *Pistacia Lentiscus*; the commercial supply has come almost exclusively from the island of Chios in the Greek Archipelago.[7]

Until the investigations of Mills regarding dammar, little was known of the chemistry of dammar and mastic, compared to the information available about rosin and shellac.[8] The resins comprise an extensive mixture of compounds; in the *Journal of the Chemical Society*, 1955, Mills and Werner list seven alcohols or ketones and four acidic constituents of the triterpene class of compounds in a sample of dammar resin. Triterpenes contain thirty carbon atoms, giving them molecular weights which range between 424 and 506. Before analyzing for the other constituents, the authors separated "dammar wax" or "β-resene," a fraction, insoluble in hot methyl alcohol, that constitutes about 29% of the original resin; this component was found to be polymeric in nature, having a low molecular weight varying from 1800 to 4000. The compounds in mastic resin were analyzed later by Seoane.[9]

When dammar or mastic is spread in thin films and exposed to the action of light, the alcoholic and ketonic compounds rapidly oxidize.[8] The acidic compounds thus formed may require that slightly different solvents be used to dissolve the aged resins; for example, these two resins are almost 100% soluble in toluene when fresh, but when aged in a carbon-arc fadeometer, they are only soluble in toluene to about 50% at room temperature. Evidence indicates that mastic oxidizes more rapidly and in this state does not dissolve as extensively in cyclohexane and toluene as oxidized dammar.[10]

Mixtures of xylene and acetone can be used as a convenient indication of the "strength" of solvent needed to remove coatings: we believe it is correct to say that neither toluene nor a 75/25 mixture of xylene and acetone will satisfactorily remove aged coatings of dammar; a 50/50 mixture, however, usually will enable an aged coating to be removed in less than thirty seconds. Such a test, similar to Ruhemann's safety margin test, is useful in "rating" coatings and is not intended to represent the solvent mixture to be used in removal from a painting.[11]

Because the natural resins are mixtures and oxidation may alter their composition in time, dissolving varnishes made with them often presents special problems. For example, a white residue is formed occasionally when an attempt is made to remove a natural resin varnish with mixtures of turpentine and acetone. Perhaps one reason for this is that not all of the components of the resin are soluble in such mixtures; it is known, for example, that a precipitate is obtained upon adding acetone to dammar-turpentine varnish.

The selection of dammar and mastic for picture varnish was based on their qualities relative to similar natural resins.[12,13,14] Shellac requires alcohol for solution, is not completely water resistant, and generally is not as pale in color. Sandarac tends to be more brittle and colophony is said to lose its gloss and discolor rapidly. Prior to the research at Mellon Institute, no extensive data on the physical properties of dammar and mastic had been published to support the preference of one over the other. The few references known contained little information suitable for making a conclusive statement regarding their properties relative to one another, or to the synthetic resins in use at the time. A minimum of data on their physical characteristics was considered necessary.[15]

It was quickly shown that dammar and mastic resins form solutions of much lower viscosity than the early synthetic resins familiar to conservators in America: Union Carbide's poly(vinyl acetate) AYAF and du Pont's poly(n-butyl methacrylate) "Lucite" ® 44 (now designated as "Elvacite" ® 2044). (The viscosity of solutions containing 20% by weight of various resins may be compared in Table 5-1.) The viscosity of solutions of

TABLE 5-1. VISCOSITY GRADE OF VARIOUS RESINS IN TOLUENE SOLUTION

RESIN	VISCOSITY IN CENTIPOISES AT 70° F. 20% SOLIDS
Dammar	1.3
Mastic	1.8
AW-2	1.2
AYAB	9
AYAA	40
AYAF	80
AYAT	167
Elvacite ® 2044	48
Elvacite ® 2045	55
Acryloid ® B-72	29
Acryloid ® B-67	18
Acryloid ® F-10	27

AW-2, Resin based on polycyclohexanone, Badische Anilin and Soda Fabrik, Germany, kindly supplied by Dr. A. E. A. Werner, London, and by the manufacturer.

AYAB to AYAT, Poly(vinyl acetate), Bakelite Division, Union Carbon and Carbide Corporation.

Elvacite ® 2044, Poly(n-butyl methacrylate), Elvacite ® 2045, poly(isobutyl methacrylate), Polychemicals Department, E. I. du Pont de Nemours & Co., Inc.

Acryloid ® B-72, a methacrylate co-polymer, Acryloid ® B-67, and Acryloid ® F-10, methacrylate polymers, Resinous Products Division, Rohm and Haas Co.

the best grade dammar and mastic resins in toluene is given in Figure 5-1; the crosses refer to previous data on dammar resin determined by Mantell and Skett.[16] These authors also point out that aromatic compounds in petroleum strongly reduce the viscosity of dammar solutions.

The hardness of dried films of dammar varnish is nearly the same as that of mastic and, on the Sward Hardness scale, greater than those of poly(vinyl acetate) and poly(n-butyl methacrylate), as shown in Table 5-4.[15] In this measurement, a rocking device is set in motion on the surface of the film; soft films dampen the motion of the Sward Rocker more rapidly than harder ones.

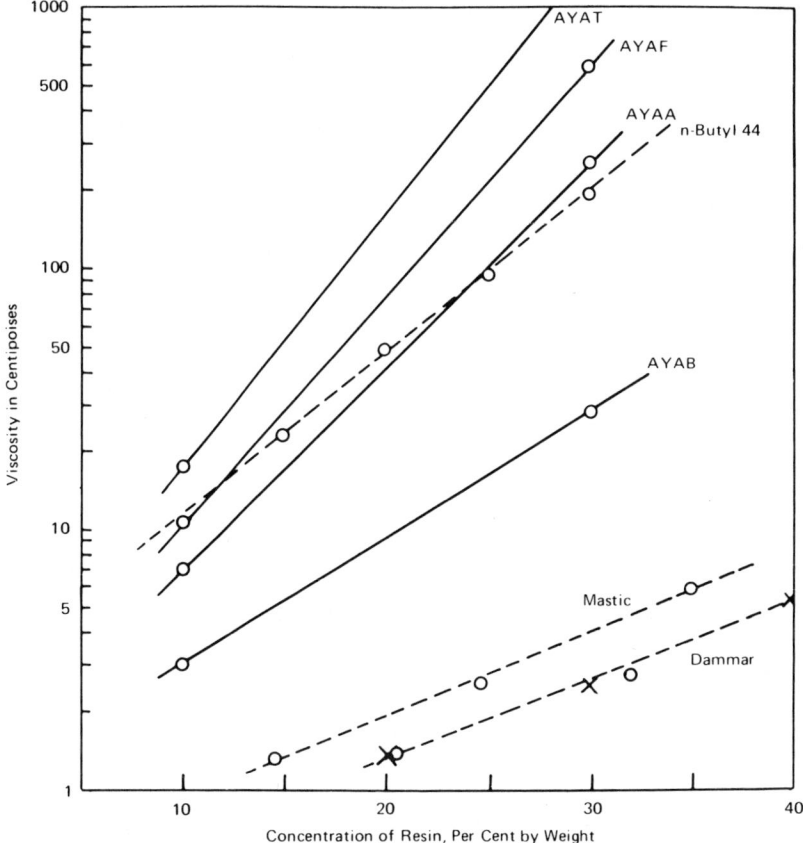

*Figure 5-1   Viscosity in Toluene at 70°F.
(n-Butyl 44 is now designated Elvacite ®2044)*

The brittleness of the films was measured by determining the diameter of mandrel around which they fail to withstand bending, when coated on a sheet of aluminum foil. Comparative data in Table 5-2 indicate that the dried dammar and mastic coatings are highly brittle. The addition of 5% stand oil by volume to dammar varnish (35% solids by weight), which is often done, does not appreciably change the brittleness of thoroughly aged films.[15]

The infra-red spectra of dammar resin was reported first by Hanson. Shortly thereafter, our laboratory published a comparison with that of mastic.[17,18] Since then Thomson and Masschelein-Kleiner have published extensive studies of the infra-red absorption of these and other resins of interest.[19,20] Houwink has reported their moduli of elasticity.[21]

TABLE 5-2. CRACKING OF BAKED FILMS OF DIFFERENT VISCOSITY GRADE RESINS MEASURED AT 70°F. (21°C.), 50% R.H.*

RESIN	VISCOSITY GRADE, CENTIPOISES	DIAMETER OF MANDREL, IN INCHES, NECESSARY FOR CRACKING
AW-2	1.2	4.0
Dammar, Mastic	1.5	2.7
Polystyrene	12.0	1.3
Poly(vinyl acetate) AYAB	9.0	0.6
Poly(vinyl acetate) AYAA	40.0	<0.1
Poly(isoamyl methacrylate)	8.1	0.4
Poly(isoamyl methacrylate)	9.6	0.3
Poly(isoamyl methacrylate)	15.2	0.2
Poly(isoamyl methacrylate)	22.2	0.1
Poly(isoamyl methacrylate)	30.2	<0.1
Poly(n-butyl methacrylate) Elvacite ® 2044	48.0	<0.1
Poly(isobutyl methacrylate) Elvacite ® 2045	55.0	<0.1
Acryloid ® B-72	29.0	<0.1

*Values for 0.0015″-thick films coated on 0.001″-thick aluminum foil. Films cast from toluene solutions of the resins were allowed to stand for one week at room temperature, then were baked for two days at 70°C., and allowed to stand for ten days at constant temperature and humidity before testing. The baking was intended to drive off most of the solvent, but not to cause marked deterioration of the resin.

One of the major objections to dammar and mastic picture varnishes has been their tendency to discolor. Thus far few attempts have been made to identify the origin of this discoloration, but the presence of turpentine probably does nothing to improve the situation. Strong sunlight or ultra-violet light will bleach the color to some degree, but the discoloration of these traditional varnishes on pictures in art galleries cannot be denied; an entire generation at one time had become accustomed to an amber "gallery tone" in paintings.[22] Dammar and mastic varnishes also present an annoying problem of *bloom*, the tendency to develop an obscuring fogginess. In addition, the traditional methods of removing these varnishes from paintings often necessitates the use of solvents that are not among those that exhibit the least tendency to act on oil paint.

It is not our intention to list all the advantages and disadvantages of dammar and mastic resins. Moreover, the traditional choice of these two resins over other natural resins should not be taken by the research chemist to mean that the development of a new picture varnish should seek to duplicate particular properties of these resins, such as their hardness or their characteristics of viscosity in solution. The vast variety of resins available today through controlled chemical synthesis suggests the possibility that materials for art and conservation may be improved and expanded. If the properties required of a picture varnish can be established and properly defined, it should be possible to find a suitable resin to fulfill the specifications; we will say more about what these properties might be in Chapter 7. This section, however, has described a sufficient number of the properties of dammar and mastic resins so that they can be compared objectively with synthetic resins.

SYNTHETIC THERMOPLASTIC POLYMERS

Synthetic resins are of recent origin. Since the industrial development of cellulose nitrate and acetate (1900 to 1910), phenol formaldehyde (1909), poly(vinyl acetate) (1928), and acrylics (1931), the number and types of synthetic resins have grown overwhelmingly. The subject can be simplified by considering that there are two major physical classifications:

1. *Thermosetting* resins which, upon heating, become resistant to solution, hard, and infusible; and
2. *Thermoplastic* resins which, upon repeated heating and cooling, retain their ability to soften under heat and to dissolve. Thermoplastics are used in solvent-type varnishes.

Almost from the time they first became available, synthetic thermoplastic resins have attracted attention in the care of objects of art. Lucas early described uses for nitrocellulose, although it is now little used because of its inherent instability.[23] At an international conference at Rome in 1930, Stout described the properties of poly(vinyl acetate), and information concerning its properties and uses was published soon after.[24,25,26] Considerably less has been published about the first applications of methacrylate polymers in conservation. Experiments with these began at the Fogg Art Museum at Harvard in the early thirties; the types long used for protecting pictures have been mainly polymers of normal-butyl and isobutyl methacrylate.[27,28,29] In 1963, attention was drawn to the excellent properties of the Rohm and Haas polymer Acryloid ® B-72, a co-polymer of ethyl methacrylate and methyl acrylate.[30]

In the two decades prior to 1959, about eight or nine proprietary picture varnishes based on methacrylate and vinyl acetate polymers were made available to artists in America. In the decade since then, the number here and abroad has increased markedly; dealers in artists' supplies now offer a variety of pressurized containers designed for immediate spray-application of natural and synthetic resins as general-purpose protective coatings. However, the synthetic resins used in proprietary formulations are seldom fully specified in labeling or advertising; museum conservators have tended to obtain the polymers directly from manufacturers and to formulate their own coatings of known composition.

Linear, long-chain polymers are created by causing reactive molecules of low molecular weight (monomer) to join together in a continuous chain (polymer). The monomer usually is a mobile liquid (styrene, molecular weight 104, boiling point 145-146°C.; vinyl acetate, m.w. 86, b.p. 72-73°C.; n-butyl methacrylate, m.w. 142, b.p. 163-164°C.). These individual molecules can be joined to form a chain having a molecular weight in the thousands, much as paper clips (monomer) might be joined to form a chain (polymer). The number of monomer units (paper clips) in a chain is designated as the degree of polymerization (DP). The chains are not all of equal length in any given preparation; hence, an average-molecular-weight or average-degree-of-polymerization is reported. The distribution of molecular weights in several proprietary poly(vinyl acetates) of different average-molecular-weight has been reported by Blom.[31]

When a molecule of water, alcohol, or ammonia is lost in joining monomer units together, the product is known as a condensation polymer; proteins, nylon, polyesters and Resin AW-2 are of this type. When an ethylenic double bond in a monomer is merely "opened-up" and a chain caused to form, the product is called an addition polymer; acrylic and vinyl acetate polymers are of this type.

If the individual strands or chains are joined together at one or more points, a cross-linked structure is formed; this is characteristic of the thermosetting-type resin. When the chains are thus joined at various points by primary chemical bonds, the substance may swell under the influence of solvents, but the possibility of dispersion into solution is greatly diminished. The linear-chain type is more likely to be soluble.

As the average-molecular-weight of a linear polymer increases, the physical properties of the resin are altered; the softening temperature is increased, for example, and brittleness decreased. A higher degree-of-polymerization, i.e., a greater length of the polymer chains, yields a film of greater tensile strength.[32] The effect of increasing molecular weight, however, becomes less noticeable as the higher molecular weights are reached.

As the length of the polymer chains increases, the viscosity of solutions of the polymer also increases. This particular phenomenon is used to estimate the average-molecular-weight, based on a determination of the *intrinsic viscosity* of a polymer.[33]

The determination of intrinsic viscosity is a specialized measurement not likely to be made by the conservator. As a more convenient measure for characterizing resins, the Research Project has used the expression "viscosity grade"; this is defined as the viscosity in centipoises of a solution of the resin in toluene at 70°F.(21°C.) and at a concentration of resin of 20% by weight. Most resins of interest to conservators are soluble in toluene. Although the measurement at 20% solid concentration has no theoretical basis, it is easy to perform in the laboratory and is done at a resin concentration close to that found in many traditional varnishes; a similar measurement has been used to classify chlorinated rubber.[34] Table 5-1 lists viscosity grades of proprietary thermoplastic resins of interest to conservators in America. The viscosity of several of these at various concentrations in toluene is also shown in Figure 5-1.

Bending and flexing films coated on aluminum foil provide a convenient method of comparing their brittleness. If desired, the percentage elongation-at-break can be estimated from the diameter of a mandrel around which supported film must be bent in order to crack the film. Table 5-2 compares viscosity grades of various resins with their ability to withstand bending. The poly(isoamyl methacrylate) polymers and the poly(vinyl acetate) resins from the Bakelite Division, Union Carbide and Carbon Corporation, are examples of two series having increasing average-molecular-weight (the latter increase in DP from AYAB to AYAT, see Appendix E): each possesses the same refractive index, density, and almost the same hardness within the series, but their tensile strength and ability to elongate increases as the viscosity grade increases.

Knowledge of the viscosity grade is convenient in comparing polymers in a series based on the same monomer. When polymers of different types, such as polystyrene, poly(isobutyl methacrylate) and poly(vinyl acetate), possess the same viscosity grade, it does not necessarily follow that they will have precisely the same strength and brittleness: the much greater inherent brittleness of polystyrene can be seen in Table 5-2 where the brittleness of 12 centipoise polystyrene may be contrasted with 9 to 10 centipoise poly(vinyl acetate) or poly(isoamyl methacrylate). In general, however, it can be expected that low-viscosity-grade polymers will be more brittle than those of much higher viscosity grade. For example, dammar and mastic (viscosity grade 1.5 centipoises) are much more brittle than Elvacite ® 2044 (48 centipoises) or resin AYAF (80 centipoises). The relationship of viscosity grade to brittleness is clearly illustrated in Figure 5-2.[30]

Because of the various viscosity grades available and the resulting differences in their physical properties, the polymers used in conservation should be specified precisely. In America, Bakelite "vinyl" resin AYAF has been the variety of poly(vinyl acetate) frequently used; the methacrylates Elvacite ® 2044 and 2045 have been widely used in the treatment of paintings. Acryloid ® B-72 was applied early to metal objects.[35] Poly(vinyl acetate) has been used extensively as both an adhesive and a protective coating.

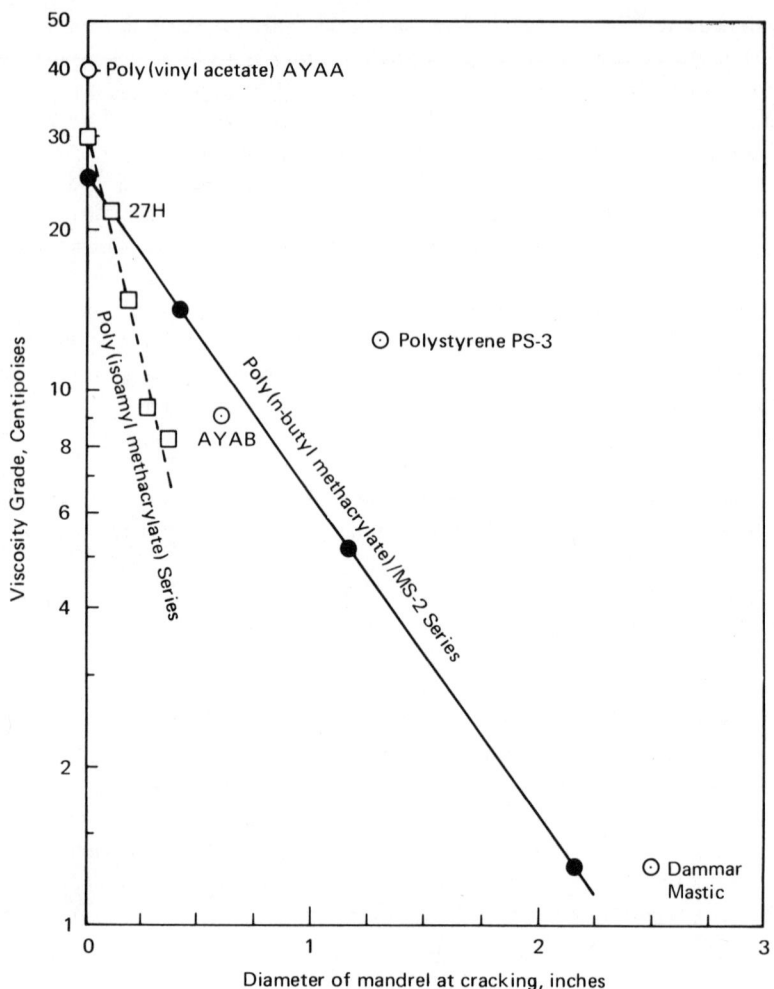

Figure 5-2 Brittleness of Films at 70°F. and 50 Per Cent Relative Humidity as a Function of Their Viscosity Grade (1.5-mil-thick films on 1-mil aluminum foil).[30]

Synthetic resins do not necessarily possess high viscosity grades, but cover a wide range. Badische Anilin and Soda Fabrik's Resin AW-2, a resin in the range of viscosity similar to dammar, has long been under investigation at the National Gallery, London.[27,36,37] This resin, based on polycyclohexanone, was also available from Howards of Ilford, England (Howards and Sons, Ltd.,

Canada) and was designated Resin MS-2. Without an added plasticizer the resin is extremely brittle (see Table 5-2). Hydrogenated forms known as MS-2A and MS-2B have been produced, but all of these proprietary resins have been withdrawn from the market at one time or another.[38,39,40] Inquiries regarding their properties and availability are best directed to Mr. Garry Thomson at the National Gallery, Trafalgar Square, London, W.C.2, who has studied their properties extensively. Straub has described their use in retouching.[41]

Few synthetic polymers sold for the purpose of forming films, intended to exhibit good tensile strength and elongation-at-break when used alone without the addition of non-volatile plasticizers, have viscosity grades less than 20 centipoises. The viscosities of 20% solutions of several well-known commercial resins are given in Table 5-3 to illustrate this point. It can be seen

TABLE 5-3. VISCOSITY OF SOLUTIONS OF VARIOUS COMMERCIAL THERMOPLASTICS AT A CONCENTRATION OF 20% BY WEIGHT

RESIN	VISCOSITY IN CENTIPOISES	TEMPERATURE °C.	SOLVENT
18-25 Centipoise RS Nitrocellulose (Hercules) DP <37*	125	25	Butyl acetate-Ethyl alcohol
1/4 Second RS Nitrocellulose 37 <DP < 60*	400	25	Butyl acetate-Ethyl alcohol
Poly(n-butyl methacrylate) (du Pont)	48	21	Toluene
Poly(ethyl methacrylate) (du Pont)	140	21	Toluene
Poly(vinyl chloride-acetate) VYHH (Bakelite)	100	21	Methyl ethyl ketone
Acryloid ®F-10 (Rohm and Haas)	27	21	Toluene
Acryloid ®B-67 (Rohm and Haas)	18	21	Toluene
Acryloid ®B-72 (Rohm and Haas)	29	21	Toluene

*W. A. Caldwell and J. J. Creasy, *J. Oil & Colour Chemists' Assoc.* 37 (1954): 63-83.

that the Acryloid ® polymers are prepared at viscosity grades close to the minimum that will provide satisfactory film strength. Figure 5-2, showing the relationship of viscosity grade to brittleness, can be used conveniently to compare the brittleness of any polymer system that may be considered as a protective coating for museum objects.[30]

The limited number of polymers adapted by conservators since 1930 have a common characteristic which may not be apparent at first: they possess reasonable flexibility without the need of added plasticizers, which are substances added to coating materials to make them more pliable; they may be relatively *non-volatile* or *volatile*. If they are the non-volatile type, they should properly be considered as part of the non-volatile component of a solvent-type varnish. Many plasticizers commonly used in industrial coatings, however, are relatively volatile; they need to retain their beneficial effect only throughout the expected lifetime of a toy or other disposable object. Dibutyl and dioctyl phthalate are typical examples. Early investigators who sought to adapt synthetic resins for picture varnish avoided the use of volatile plasticizers because many have the ability to soften traditional paints and varnishes. Therefore, in order to avoid the possibility that the plasticizer might migrate into the painting and to obtain long-term stability of properties, the tendency has been to select polymers which are pliable in themselves for the care of objects of art. It has been said that polymers such as poly(n-butyl methacrylate) are *internally plasticized*; their inherent flexibility will last as long as the polymer itself withstands the forces of time. Note the decrease in hardness in the series from the ethyl through n-propyl to the n-butyl methacrylate polymer (Table 5-4): the substitution of an n-butyl group for the ethyl or n-propyl effectively softens the polymer.

TABLE 5-4. HARDNESS OF THERMOPLASTIC RESINS*

RESIN	VISCOSITY GRADE, CENTIPOISES	SWARD HARDNESS	PENCIL HARDNESS
Poly(n-butyl methacrylate)	27	–	3B-2B (Soft)
Poly(n-butyl methacrylate)	50	30	2B
Poly(isoamyl methacrylate)	22	49	B-HB
Poly(n-propyl methacrylate)	20	–	HB-F
Poly(vinyl acetate)	9	63	F
Poly(vinyl acetate)	40	–	F-H
Poly(vinyl acetate)	80	–	H
Poly(isobutyl methacrylate)	55	65	H
Poly(ethyl methacrylate)	140	–	2H-3H
Dammar, Mastic	1.5	81	(Hard)

*Films 0.0015" thick, baked on window glass (1/16" thickness) and measured at 70°F. (21°C.) and 50% relative humidity.

To summarize the foregoing remarks before continuing: linear polymers have been described as atoms or groups of atoms joined together to form chain-, ribbon-, spaghetti-like molecules. The fundamental significance of the average length of the chain, and whether or not the chains may be cross-linked has been pointed out. Thus, one may consider that a vast majority of polymers fall into two major classifications based upon chemical structure: 1) cross-linked, and 2) linear, long-chain. The closely related designations 1) thermosetting, and 2) thermoplastic, are based upon differences in physical behavior.

Bearing in mind the general designations of thermoplastic or thermosetting, it is obvious that polymers may also be classified according to their chemical constitution. The polystyrene class is based on the monomer, styrene. Polymers based on vinyl chloride, vinyl acetate, and vinyl alcohol are familiar. In polymethacrylates, the particular ester of the methacrylic acid involved must be specified. The chemical nature of the monomer influences such properties of the polymer as its tendency to dissolve in a given solvent, hardness, and durability.

Resins differ in the type of solvents required to dissolve them. As a convenient measure of the relative tendencies for resins to dissolve in hydrocarbon solvents, *solubility grade* has been defined (see Chapter 3, p. 34). In the future, sufficient data and experience may be available to permit classification of polymers according to their solubility parameter or *cohesive energy density*.

Hardness is a property that, although readily sensed, has never been defined satisfactorily. Customarily, it is defined only in terms of the apparatus or method of measurement, e.g., indentation hardness, penetration hardness, scratch hardness, rocker hardness; each of these measures something slightly different. A comparative list of the hardness of familiar resins is given in Table 5-4. The very simple measurement of the pencil-hardness test[42] distinguishes many of the polymers encountered and may be correlated with the more precise Sward hardness. The step-wise spacing of the pencil scale for the Eagle "Turquoise" pencils changes at F.[43]

Figure 5-3 presents a convenient chart which summarizes the information just presented.[44] It can be seen that the properties of resins may be classified conveniently in terms of three parameters: viscosity grade, solubility, and hardness. Although the latter property is not well-defined, it is related to the modulus of elasticity.

Protective coatings differ in their resistance to deterioration. At present, there is considerable knowledge of this phase of polymer chemistry, and chemists are able to choose chemical structures that are expected to be resistant to reaction with oxygen or other reactive agents. For example, the energy required to break the bonds in silicones and fluorohydrocarbons is high; they have been shown to possess great durability under severe conditions, and may someday find application as permanent artists' materials. Many of these compounds, however, present special problems in adhesion and

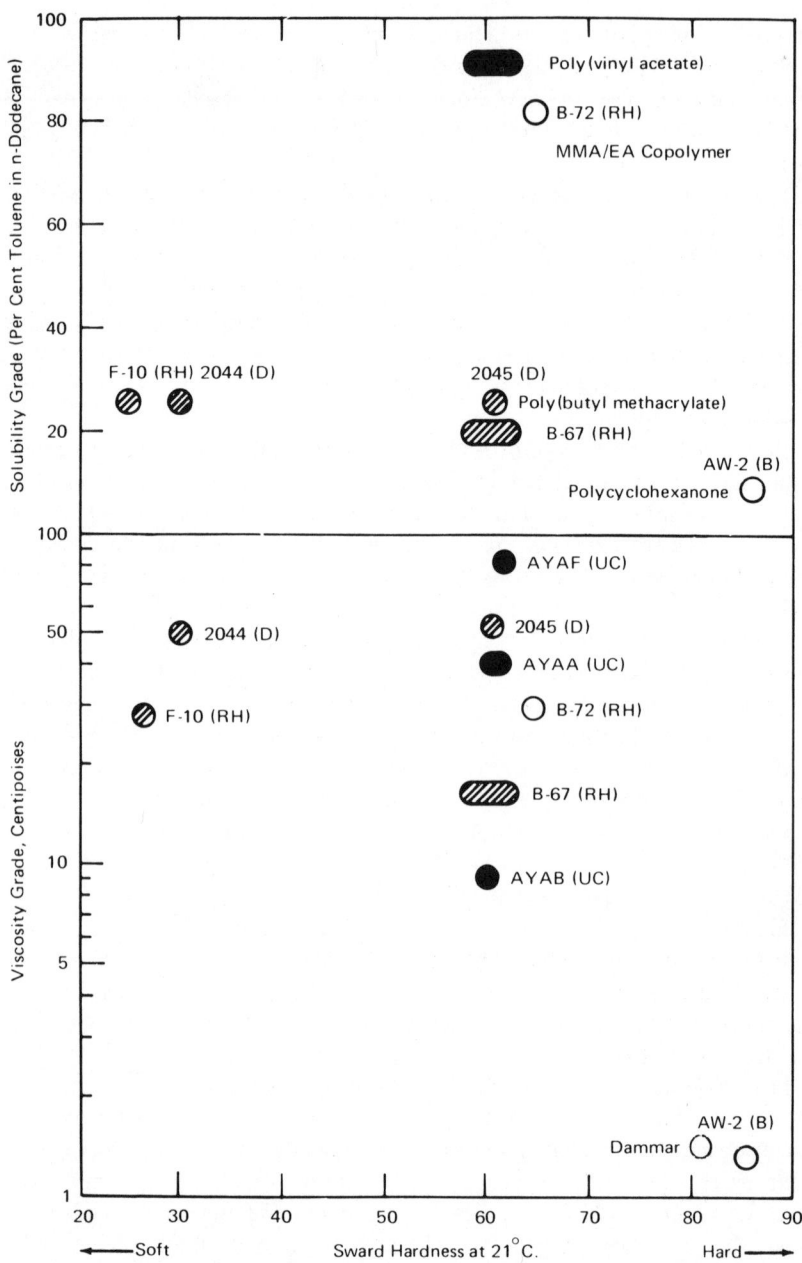

Figure 5-3 Properties of Familiar Synthetic Resins.
(UC) Union Carbide (D) du Pont (RH) Rohm & Haas (B) BASF. Based on Data in Table 2 in Ref. 30.

solubility which rule out their consideration for picture varnish at the present time. Polystyrene and poly(vinylidene chloride) present problems with regard to their stability to light.[45] Poly(vinyl acetate)[46] and polymethacrylates[47] have been shown to be highly resistant to deterioration at normal temperatures and for this reason they have attracted much attention in conservation. The tendency of certain methacrylate polymers to cross-link is now well understood and types that do not exhibit this tendency can be selected. Resins AW-2, MS-2A, and related polycyclohexanone resins appear to be resistant to oxidation; a study of their properties continues.[19]

The refractive index and permeability of a material are determined partly by the atoms of which it is constituted; hence, these properties will vary among polymers of different chemical composition. The limited role played by refractive index in picture varnish has been discussed elsewhere.[19,48] Permeability will be considered in Chapter 7.

In short, the characteristics of synthetic resins vary over a wide range. One cannot recommend or condemn the use of synthetic resins in the fine arts as a class, but must consider the merits of each one individually for a given application. Even such designations as poly(vinyl acetate) and "acrylic polymer" must be further qualified as to the viscosity grade involved and the particular member of the methacrylate family; materials must be fully identified with regard to trade name and designation. In comparative studies of thermoplastic resins, the National Gallery of Art Research Project has found it useful to refer to the viscosity grade and hardness, as done in Figure 5-3; wherever possible solubility grade and brittleness are also determined.[49]

REFERENCES AND NOTES

1. D. C. Carman, *The Chemical Constitution and Properties of Engineering Materials* (London: E. Arnold and Company, 1919).
2. A. K. Doolittle, *The Technology of Solvents and Plasticizers* (New York: Wiley, 1954).
3. K. H. Meyer, *Natural and Synthetic High Polymers* (New York: Interscience, 1950).
4. "Reduction Products of Natural Resins, etc., for Varnishes, Lacquers, Adhesives, etc.," French Patent 804,344 (1936).
5. R. L. Feller, "What's in a Name," *The Crucible* (Pittsburgh Section, American Chemical Society) 49 (October, 1964): 214-18; "First Description of Dammar Picture Varnish Translated," *Bulletin of the American Group-IIC* 7, No. 1, (1966): 8, 10.
6. J. Plesters, "Dark Varnishes—Some Further Comments," *Burlington Magazine* 104 (1962): 452-60.
7. C. L. Mantell, et al., *Technology of Natural Resins* (New York: Wiley,

1942); C. L. Mantell and R. W. Allen, "Solubilities of Natural Resins in Solvents and Waxes," *Ind. Eng. Chem.* 30 (1938): 262-269.
8. J. S. Mills and A. E. A. Werner, "Paper Chromatography of Natural Resins," *Nature* 169 (1952): 1064; F. I. G. Rawlins and A. E. A. Werner, "Some Scientific Investigations at the National Gallery, London," *Endeavour* 13 (1954): 140-46; J. S. Mills and A. E. A. Werner, "Partition Chromatography in the Examination of Natural Resins," *J. Oil & Colour Chemists' Assoc.* 37 (1954): 131-42; "The Chemistry of Dammar Resin," *J. Chem. Soc.* (London) (1955): 3132-40; J. S. Mills, "The Constitution of the Natural Tetracyclic Triterpenes of Dammar Resins," *J. Chem. Soc.* (London) (1956): 2196-2202.
9. E. Seoane, "Further Crystalline Constituents of Gum Mastic," *J. Chem. Soc.* (London) (1956): 4158-60; E. Seoane and H. R. Barton, "The Constitution and Stereochemistry of Masticadienonic Acid," *J. Chem. Soc.* (London) (1956): 4150-57.
10. R. L. Feller, "A Note on the Exposure of Dammar and Mastic Varnishes to Fluorescent Lamps," *Bulletin of the American Group-IIC* 4, No. 2 (1964): 12; "The Deteriorating Effect of Light on Museum Objects," *Museum News, Technical Suppl.* No. 3 (June, 1964).
11. H. Ruhemann, *The Cleaning of Paintings* (New York: Praeger, 1968).
12. H. Trillich, "Dammar und Dammarlacke," *Techn. Mittheilungen für Malerei* 53 (1937): 82-84.
13. M. Toch, "Dammar as a Picture Varnish," *Technical Studies in the Field of the Fine Arts* 2 (1934): 149-52.
14. C. J. Holmes, "An Essay on Mastic Varnish," *The Burlington Magazine* 35 (1919): Part II, 68, 71, 72, 75.
15. R. L. Feller, "Dammar and Mastic Varnishes—Hardness, Brittleness, and Change in Weight Upon Drying," *Studies in Conservation* 3 (1958): 162-74.
16. C. L. Mantell and A. Skett, "Viscosities of Dammar Solutions," *Ind. Eng. Chem.* 30 (1938): 417-22.
17. N. W. Hanson, "Some Recent Developments in the Analysis of Paints and Painting Materials," *Official Digest, Federation of Paint and Varnish Production Clubs* No. 388 (1953): 163-74.
18. R. L. Feller, "Dammar and Mastic, Infrared Analysis," *Science* 120 (1954): 1069-70.
19. G. Thomson, "New Picture Varnishes," in *Recent Advances in Conservation* (London: Butterworths, 1963), pp. 176-84.
20. L. Masschelein-Kleiner, "Perspectives de la chimie des liants picturaux anciens," *Bulletin Inst. Royal du Patrimoine Art.* 6 (1963): 109-26; R. Kléber and L. Masschelein-Kleiner, "Contribution a l'analyse des com-

poses resineux utilises dans les oeuvres d'art," *Bulletin Inst. Royal du Patrimoine Art.* 7 (1964): 196-218.
21. R. Houwink, "Amorphous Substances," *Elasticity, Plasticity and Structure of Matter* (Washington, D.C.: Harron Press, 1953): Chapter 6, p. 141.
22. *An Exhibition of Cleaned Pictures* (London: The National Gallery, 1947).
23. A. Lucas, *Antiques, Their Restoration and Preservation*, 2nd ed. (London: E. Arnold and Company, 1932).
24. A. Eibner, "Neue Wege zum Oberflächenschutz," *Farben-Zeitung* 36 (1931): 1849-50.
25. R. J. Gettens, "Polymerized Vinyl Acetate and Related Compounds in the Restoration of Objects of Art," *Technical Studies in the Field of the Fine Arts* 4 (1935): 15-27.
26. G. L. Stout and R. J. Gettens, "Transport des fresques orientales sur de nouveaux supports," *Mouseion* 17 (1932): 107-12.
27. A. E. A. Werner, "Plastics Aid in Conservation of Old Paintings," *British Plastics* 25 (1952): 363-66.
28. C. K. Keck, *How to Take Care of Your Pictures* (New York: The Museum of Modern Art and the Brooklyn Museum, 1954), p. 24; S. Keck, "The Care and Cleaning of Your Picture," *The Brooklyn Museum Bulletin* 10, No. 3 (1949): 1-12.
29. R. J. Quandt, "Reclamation of Two Paintings," *The Corcoran Gallery of Art Bulletin* 6, No. 3 (1953): 2-15.
30. R. L. Feller, "New Solvent-type Varnishes," in *Recent Advances in Conservation* (London: Butterworths, 1963), pp. 171-75.
31. A. V. Blom, "Synthetic Film-forming Materials: Vinyl and Allied Polymers," *Organic Coatings* (New York: Elsevier, 1949), p. 101.
32. D. C. Carman, "Properties of High Polymers, Viscosity," *Chemical Constitution and Properties*, Chapter XIX, p. 588; H. F. Payne, "Fundamentals of Film Formation, Lacquer Films," *Organic Coating Technology* (New York: Wiley, 1954), Chapter 1, p. 36.
33. K. H. Meyer, "I. Investigation in Solution. Characterization by Means of Solvents and Swelling Agents," *Natural and Synthetic High Polymers*, p. 32; also "Determination of Molecular Weight from Measurements of Viscosity," *ibid.*
34. A. V. Blom, "Materials from Natural Sources, Rubber Derivatives," *Organic Coatings*, p. 72.
35. "Acryloid Helps Preserve Art Treasures," *The Rohm and Haas Reporter* 8, No. 3 (1950): 14.
36. A. E. A. Werner, *The Scientific Examination of Paintings*, Lectures, Monographs, and Reports No. 4 (London: Royal Institute of Chemistry, 1952).

37. R. L. Feller, "Characteristics of Resin AW-2," *Bulletin of the American Group-IIC*, 3, No. 1 (1962): 9-10.
38. G. Thomson and N. S. Brommelle, "Changes in Manufactured Products," *Museums J.* 64 (1964): 82-83; "Changes in Manufactured Products," *IIC-News* 3, No. 1 (1964): 9.
39. H. Lank, "MS2B," *IIC News* 3, No. 3 (1965): 6.
40. M. Hey, "MS2A Picture Varnish Resin," *IIC News* 4, No. 4 (1967): 8.
41. R. Straub, "Retouching with Synthetic Resin Paint," *Museums J.* 62 (1962): 113-19.
42. S. D. Coleman and E. N. Smith, "The Pencil-Hardness Test for Organic Finishes," *Organic Finishing* 10 (1949): 23.
43. Eagle Pencil Company, New York (private communication).
44. R. L. Feller, "Research on Durable Thermoplastic Polymers for the Conservation of Works of Art," *Atti della XLIX Riunione SIPS, Siena*, 23-27 September 1967 (Rome, 1968): 1099-1110.
45. L. A. Matheson and R. F. Boyer, "Light Stability of Polystyrene and Polyvinylidene Chloride," *Ind. Eng. Chem.* 44 (1952): 867-74.
46. *Gelva* (New York: Shawinigan Products Corp., 1948); *Vinylite: Vinyl Acetate Resins* (New York: Union Carbide and Carbon Corp., Bakelite Division, 1951).
47. B. M. Axilrod and G. M. Kline, "Study of Transparent Plastics for Use on Aircraft," *Nat. Bur. Standards, J. Research* 19 (1937): 367-400.
48. R. L. Feller, "Factors Affecting the Appearance of Picture Varnish," *Science* 125 (1957): 1143-44.
49. Unless otherwise noted, the du Pont products Elvacite ®2044 and 2045 are the poly(n-butyl methacrylate) and poly(isobutyl methacrylate) polymers, respectively, referred to in the text. The trade designations of the polymers exposed at the Fogg Museum of Art are noted in Table 7-5, p. 159, and on p. 172.

6
PROPERTIES OF FRESHLY APPLIED VARNISH

DRYING OF VARNISH

Many of the details that should be considered under this title have already been discussed in earlier chapters. We shall recall the pertinent factors and organize them in such a manner that the drying of solvent-type varnish and the control of its appearance may be discussed as specific topics. This chapter begins with the statement that a solvent-type varnish dries by loss of solvent and becomes dry to the touch long before the solvent has departed; typical drying curves were given in Figure 1-1. As much as 15% of the weight of the film of dammar-turpentine varnish represents solvent after it has dried for one month at 70°F. and 50% relative humidity.[1] When toluene is the solvent, about 8% is retained after a month. Traces of solvent usually linger for long periods of time; chemists in the industrial-coatings field have emphasized repeatedly the difficulty of removing the last traces of solvent from films, especially with solvents that "associate," or exhibit particular compatibility, with the resin.

If the solvent is easily oxidized, the film may deteriorate faster than otherwise. Dammar varnish formulated with old turpentine exhibits a marked tendency to oxidize, whereas varnish prepared with freshly distilled turpentine does not.[1] If the retained solvent discolors, it may lead to discoloration of the coating even though the resin itself is highly resistant. (In Chapter 2, the substitution of cellosolve acetate for diacetone alcohol in poly(vinyl acetate) coatings was cited, being done partially to avoid the yellowing tendencies of the latter solvent.)

The rate of drying, although influenced in part by the nature of the nonvolatile components of a film,[2] is usually controlled by choosing solvents which evaporate at greater or lesser speeds. Figure 6-1 illustrates the effect of solvents of different evaporating rates on the brittleness of dammar and mastic varnishes after one month of drying.[1]

138   RESINS AND THE PROPERTIES OF VARNISHES

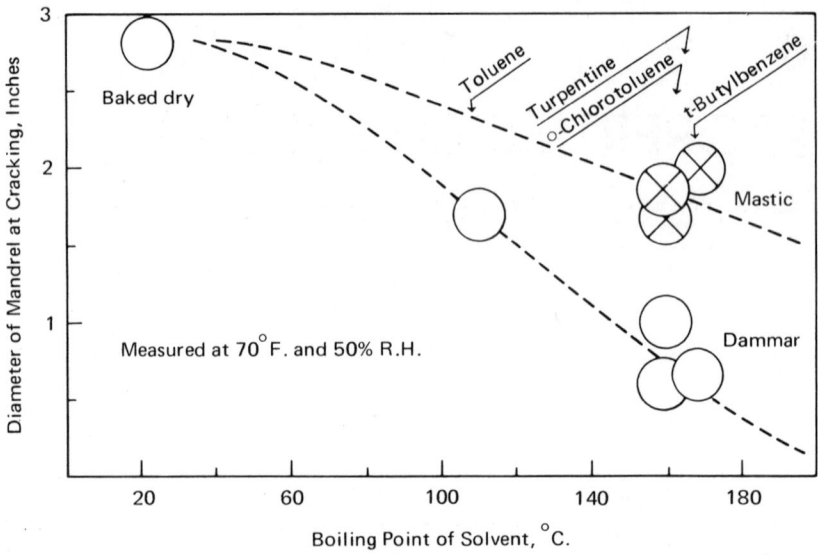

Figure 6-1 Influence of Solvent Upon Brittleness: 1.5 mil. Films of Dammar and Mastic After One Month.

In the early minutes of drying, the loss of solvent causes the concentration of solids to increase, with the result that the viscosity of the varnish increases. The change that takes place with a solution of poly(vinyl acetate) AYAF in toluene formulated at 10% solids concentration may be followed in Figure 5-1: as the toluene evaporates, the viscosity will climb along the line of the graph. In a very short time, after only about one-fourth of the solvent has evaporated, the solution will attain a viscosity greater than 1000 centipoises. A rapid increase in viscosity such as this can result in rough-textured coatings, as well as stringing and pulling of the brush, or webbing from a spray gun.[3,4,5] In order to avoid these difficulties, it is frequently advantageous to add slow-evaporating solvents in formulating varnish. Another way to minimize the effects of high viscosity, of course, is to employ resins that have low viscosity grades; one simply selects a resin with the lowest viscosity grade that will still provide films of satisfactory strength and toughness (Figure 5-2).

The successful formulation of solvents to achieve proper gloss and satisfactory drying rates is a subject discussed extensively in the industrial literature on surface coatings. In order to reduce costs in industrial applications, a mixture of solvents is frequently used, a practice that introduces many technical problems. For example, if the least-rapid evaporating component of a mixture of solvents is also a poor solvent for the resinous material, then, as the rapidly evaporating components depart, the resin may come out of solution, resulting in a coating of poor physical structure. Similarly, if the fastest evaporating components of the solvent are compatible with water, then the water may separate from the formulation as the solvent evaporates, again causing unsatisfactory appearance. The industrial literature speaks of "balanced" formulations of solvents; the term refers in part to the balance of the evaporating rates of both the good- and bad-solvent components in a mixture.[5]

Chapter 3 mentions another way of classifying solvents, that is, by the viscosity of their solutions with a particular resin. In many formulations, a solvent is frequently sought that gives the minimum viscosity to a varnish. (The effect of aromatic compounds on the viscosity of solutions of dammar resin has been noted, p. 122.)

To summarize the broad aspects of the subject of drying: solvent-type varnishes lose their solvent rapidly at first, but require a long time before the last significant traces are gone; Figure 1-1 illustrates the manner in which the solvent is lost from several typical coatings. Traces of solvent that are present in the film after it is dry to the touch influence such properties as hardness, brittleness, and tendency to yellow. The rate at which solvent evaporates during application of a varnish markedly influences its handling qualities and final appearance. Moreover, if mixtures of solvents are used in which one component may have decidedly less "solvent power" for the resin than another, their evaporating rates relative to one another will influence the final condition of the film.

CONTROL OF APPEARANCE

Discussion of the appearance of picture varnish logically follows the above remarks because the manner in which a varnish dries usually has a decided influence on its appearance. First, however, factors that are involved in the appearance of a surface coating will be considered.

Judd suggests seven factors that influence surface-coating appearance.[6,7] Surface texture and color are two of these. The other five are types of gloss; for example, specular gloss, distinctness-of-image gloss, contrast gloss, sheen, and bloom. Specular gloss refers to the reflection that occurs at the angle of reflection equal to the angle of incidence of a beam of light. Contrast gloss refers to the ratio of intensity of light that is reflected at two different angles relative to the surface. Distinctness-of-image gloss refers to the distinctness of patterns of light reflected from a surface. Frequently the mullions in a window are used in a simple test of the relative distinctness-of-image of different coatings. Sheen is a type of reflection possessed by velvet, wherein a matte surface gives a pronounced specular reflection when examined at a small angle to the surface. Bloom, as used in this sense, is distinct from the more common meaning of the word signifying a varnish defect; here it designates a halo or halation of the image in surfaces of high gloss. Thus, we may distinguish five kinds of gloss, each of which refers to a phenomenon that cannot be described by any of the others. In certain problems, it may be possible to define other types of gloss, but at present, it is commonly conceded that at least these five are necessary to describe distinct aspects of appearance.

It may readily be seen that, if a varnish dries with a rough surface, its contrast gloss and distinctness-of-image gloss will be markedly affected. Bruxelles and Mahlman supplement their remarks with a diagram, similar to Figure 6-2. Their diagram shows that, if a solvent-type varnish forms an immobile gel at a point when considerable solvent still remains, it will tend to form a surface which will follow the irregularities of the paint underneath the

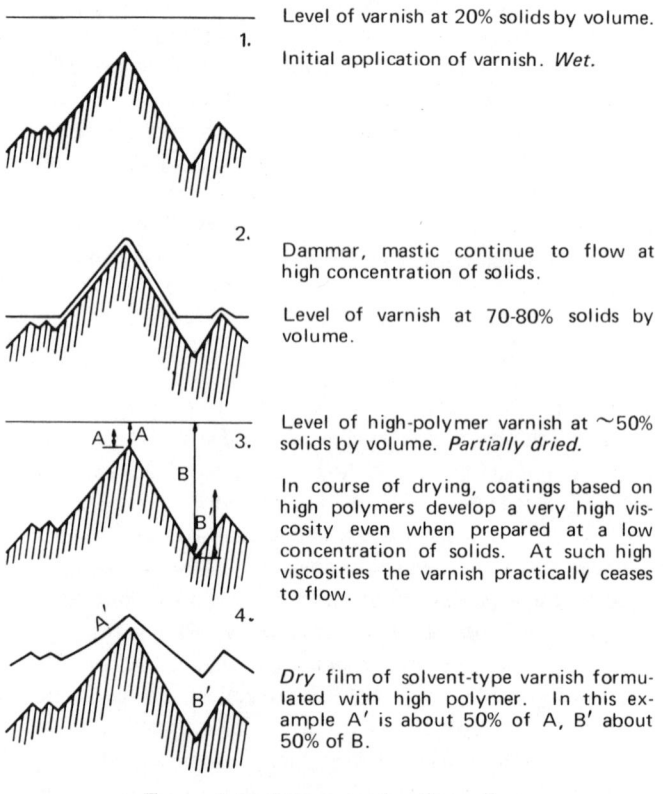

1. Level of varnish at 20% solids by volume.

   Initial application of varnish. *Wet.*

2. Dammar, mastic continue to flow at high concentration of solids.

   Level of varnish at 70-80% solids by volume.

3. Level of high-polymer varnish at ~50% solids by volume. *Partially dried.*

   In course of drying, coatings based on high polymers develop a very high viscosity even when prepared at a low concentration of solids. At such high viscosities the varnish practically ceases to flow.

4. *Dry* film of solvent-type varnish formulated with high polymer. In this example A' is about 50% of A, B' about 50% of B.

Figure 6-2 Diagram After Bruxelles and Mahlman.

varnish.[3] In like manner, a picture varnish that is formulated with polymers of high viscosity grade may rapidly become immobile owing to the rise in viscosity, and thus tend to be less glossy than those formed with resins of low viscosity grade, such as dammar, mastic or Resin AW-2. Hence, the coatings based on high polymers tend to give semi-matte finishes, often considered attractive on many modern paintings and Italian primitives in tempera. Moreover, in the repair of minor scratches and breaks in a film of high polymer, it is often effective simply to moisten the area with solvent and to allow the film to "reform," whereas dammar and mastic coatings tend to flow and form a glossy ring at the edge of the treated area.

One way to achieve as much leveling as possible in a varnish formulated with a high polymer is to slow down the rate of evaporation of the solvent; this will reduce the rate at which the viscosity climbs to that point at which the varnish will no longer flow freely. By this means, flow is allowed to take place as long as possible and a higher gloss is attained.

A coating which is not able to level itself completely forms a non-glossy surface; the result is a matte or dull finish. One of the common techniques in spray application is to choose solvents that evaporate rapidly. By holding the spray gun away from the painting in such a manner that the varnish is viscous by the time it reaches the surface of the painting, the varnish loses much of its solvent. Pomerantz has mentioned some of these facets of varnishing in his booklet for artists and collectors.[8] A formal discussion of varnishes and their appearance (including the suggestion of dulling by spray gun) may be found in the 1940 *Manual on the Conservation of Painting*, and in Ruhemann's recent book.[9,10] The subject of appearance is also discussed briefly in a recent UNESCO publication.[11]

The formulation of solvent becomes critical when resins of high viscosity grade are used in solvent-type varnishes. For example, a formulation of 2 gms. of Union Carbide's poly(vinyl acetate) AYAF (80-centipoise viscosity grade), 250 ml. of ethyl alcohol, 100 ml. of cellosolve acetate and 35 ml. of diacetone alcohol, gives a rough surface when sprayed on window glass at a distance of from ten to twenty inches from the glass. When cellosolve acetate is eliminated and a total of 135 ml. of diacetone alcohol is used in this formulation, to achieve a slower rate of drying, the distinctness-of-image gloss is markedly improved with the gun held at a distance of twenty inches from the glass.[12]

Again, it should be emphasized that the selection of solvents for picture varnish is restricted. For solvents which may have some action on paints and glazes, we may not wish to employ the slower evaporating members of the same chemical class, because the longer the solvent remains in contact with the paint, the greater the risk of solvent action. An additional complication in selecting slower evaporating solvents is discussed at length in Chapter 2: as larger molecules in a homologous series are selected, the "solvent power" does not remain constant.

In this brief discussion of how the evaporating rate controls the final appearance, it would be wrong to imply that the alteration of the evaporation rate is a simple solution to particular problems or one that may not require a compromise between alternatives. Considerable experience is necessary for the skillful modification of solvent formulations.

Another aspect of appearance that deserves mention is the fact that both the chroma (purity) and the value of colors are frequently observed to change when a painting is varnished; an important function of a varnish is to "allow the colors to be seen." A painting having an age-roughened surface exhibits considerable diffuse reflection from the surface of the paint. Such a picture requires a coat of varnish to allow the colors to be seen without the interfering scattered light. The refractive index has something to do with the success in revealing the colors, but there is, in addition, the influence of surface gloss, wetting, and penetration, which is of particular importance. Penetration into porous paint is characteristic of varnishes prepared with polymers of low viscosity grade; the data in Table 6-1 demonstrate this point when a dark color was obtained in cases where there was good penetration.[12]

TABLE 6-1. COLOR OF LEAN POLY(VINYL ACETATE)-ULTRAMARINE PAINT WHEN VARNISHED

VARNISH RESIN	SOLVENT	COLOR	VISCOSITY GRADE, CPS.	REFRACTION INDEX OF RESIN
Polymethacrylate	Cycloparaffins	Light	22	1.48
Polymethacrylate	Cycloparaffins	Dark	8	1.48
AW-2	Cycloparaffins	Dark	1.2	1.52
Dammar	Turpentine	Dark	1.3	1.53
Poly(vinyl acetate) AYAB	Toluene	Dark	9	1.46
Poly(vinyl acetate) AYAT	Toluene	Light	114	1.46
Poly(vinyl alcohol)	Water	Light	~400*	1.51

*In water

The wetting of a paint surface is related to the problem of chemical compatibility. To illustrate this point, we take an extreme case in which a paint made with pigment and wax is used. This paint is not likely to reveal the colors much more brilliantly in hue and value when coated with water because the water will not wet and form an intimate contact with the paint. The subject of wetting is mentioned because of observations of certain experimental coatings formulated in paraffin hydrocarbons at Mellon Institute:

this type of coating frequently does not satisfactorily wet horny bits of old resins that may remain on the surface of an old painting. A solvent as polar as xylene or toluene is often necessary to secure adequate wetting of the varnish on the surface of the aged painting; otherwise, tiny areas of air bubbles occasionally appear where poor contact has been made. Partly for this reason, we have proposed that as a standard of the maximum solubility parameter or "strength" of solvent to be used, polymers be sought that can be applied and removed in xylene.[13] Of course, applications can also be found in which the use of resins soluble in "milder" solvents is appropriate and effective.

When synthetic resins were first used as picture varnishes, there was some discussion concerning whether their occasional poor appearance was due to their lower refractive index in comparison to dammar and mastic.[14] So many other factors exist that usually exert a greater influence on appearance, however, that it does not seem possible that the slight difference in refractive indices can be the cause of the significant differences that are frequently observed (see Table 6-1). Marked differences in appearance can readily arise owing to differences in gloss and/or penetration of porous paint.[12] Comments on the possible role of refractive index in special cases have been made by Thomson.[11]

The appearance of many natural resin varnishes during the first few days after application often does not remain unchanged for very long; this must be taken into account when evaluating the appearance of freshly applied varnishes, as Thomson has pointed out.[15] Often a significant change in gloss can be due to a relatively minor physical change of the surface, perhaps related to the wrinkling of the surface of coatings made with low-molecular-weight resins, also reported by Thomson.[11,16]

The last few paragraphs have considered the role of picture varnish primarily as it affects the value and chroma (purity) of colors in a painting. However, the discoloration of aged picture varnish alters the hue of colors as well. Most affected by yellowed varnish are blues, followed by greens with the yellows and reds least altered.[17,18,19] There have been references to the "warmth" imparted when a picture is freshly varnished; while it is difficult to find adequate words to describe the special phenomena that we observe in objects of art, the use of the word "warmth" in this connection does not, in our opinion, refer to alteration of the color towards a warm hue owing to the yellowness of the varnish. We have measured the transmission spectrum of fresh dammar varnish (Figure 7-2) and found that the yellow hue of the varnish cannot be perceived in a fresh film as thin as those generally applied to a picture. The yellowing of dammar-turpentine varnish may be noticeable in the first year or two of its life; yet, the curator's judgment of a picture a day or two after it has been varnished is probably not influenced by a yellow tone.[20]

The subject of toned varnishes and whether painters in past centuries intended varnish to be toned, will not be discussed here. An excellent bibliography on this subject is suggested in Ruhemann's book and in the extended discussion of the cleaning problem that has appeared in the *Burlington Magazine*.[10,19] Johnston and Feller have called attention to a spectrophotometric technique that may be used to verify whether an intentionally colored varnish has been removed in cleaning.[21]

To summarize our consideration of factors that control appearance: it is essential first to consider the various aspects of gloss and the ways in which the surface of a varnish influences gloss. The rate of drying frequently has a marked effect upon the surface roughness of a protective coating; the effect is especially noticeable in spray application and in the use of resins of high viscosity grade. The wetting of the paint surface and the penetration of porous paint by the varnish also influence the final visual effect. In the usual application, the color of dammar and mastic varnish probably is not a major factor determining the appearance of the freshly varnished painting, unless the varnish is especially tinted. The range of refractive indices in the resins used for picture varnish is limited, and does not appear to be a major factor in determining differences in the appearance of varnishes.

REFERENCES AND NOTES

1. R. L. Feller, "Dammar and Mastic Varnishes—Hardness, Brittleness, and Change in Weight Upon Drying," *Studies in Conservation* 3 (1958): 162-74.
2. C. M. Hansen, "A Mathematical Description of Film Drying by Solvent Evaporation," *J. Oil & Colour Chemists' Assoc.* 51 (1968): 27-43.
3. G. N. Bruxelles and B. H. Mahlman, "Glossiness of Nitrocellulose Lacquer Coatings," *Official Digest, Federation of Paint and Varnish Production Clubs* No. 351 (1954): 299-314.
4. D. Wapler, "Oberflächenstruktur und Verdunstungsvorgang," *Farbe u. Lack* 59, No. 9 (1953): 352-59.
5. W. W. Reynolds, *Physical Chemistry of Petroleum Solvents* (New York: Reinhold, 1963).
6. D. B. Judd, "Physics and Psychophysics of Colorant Layers," *Color in Business, Science and Industry* (New York: Wiley, 1952), p. 304.
7. A. Dinsdale and F. Malkin, "The Measurement of Gloss with Special Reference to Ceramic Materials," *Trans. British Ceramic Soc.* 54 (1955): 94-112.

8. L. Pomerantz, *Is Your Contemporary Painting More Temporary Than You Think?* (Chicago: International Book Company, 1962).
9. International Museums Office, *Manual on the Conservation of Paintings* (Paris: International Institute for Intellectual Cooperation, 1940).
10. H. Ruhemann, *The Cleaning of Paintings* (New York: Praeger, 1968).
11. *The Conservation of Cultural Property* (Paris) UNESCO 1968, Appendix reprinting an earlier publication, *Synthetic Materials Used in the Conservation of Cultural Property* (Rome Centre, 1963).
12. R. L. Feller, "Factors Affecting the Appearance of Picture Varnish," *Science* 125 (1957): 1143-44.
13. R. L. Feller, "New Solvent-type Varnishes," in *Recent Advances in Conservation* (London: Butterworths, 1963): 171-75.
14. *Technical Chemicals for the Paint Trade* (Ilford, England: C. Howards of Ilford, Ltd., 1956), p. 94.
15. G. Thomson, "New Picture Varnishes," in *Recent Advances in Conservation* (London: Butterworths, 1963), pp. 176-84.
16. G. Thomson, "Some Picture Varnishes," *Studies in Conservation* 3 (1957): 64-79.
17. A. P. Laurie, *The Painter's Methods and Materials* (Philadelphia: Lippincott, 1926). Reprint: New York: Dover, 1968.
18. R. L. Feller, "Color Change in Oil Paintings," *Carnegie Magazine* (October, 1954): 276-79.
19. *Burlington Magazine* 54, No. 707 (1962): Editorial, "The National Gallery Cleaning Controversy," pp. 49-50; E. H. Gombrich, "Dark Varnishes," pp. 51-55; O. Kurz, "Varnishes, Tinted Varnishes and Patina," pp. 56-59; S. Rees Jones, "Science and the Art of Picture Cleaning," pp. 60-62. *Burlington Magazine* 54, No. 716 (1962): Joyce Plesters, "Dark Varnishes," pp. 452-60; Denis Mahon, "Miscellanea," pp. 460-70; J. A. van de Graff, "The Interpretation," pp. 471-75. *Burlington Magazine* 55, No. 720 (1963): E. H. Gombrich, "Controversial Methods," pp. 90-93; O. Kurz, "Time the Painter," pp. 94-97; S. Rees Jones, "The Cleaning Controversy," pp. 97-98; P. L. Jones, "Scientism and the Art of Picture Cleaning," pp. 98-103.
20. Although we are well aware that dammar varnish is yellow when viewed in a bottle several inches in diameter, if the phenomenon is checked with a spectrophotometer, we believe that it will be found, in the case of the varnishes of Gardner Color No. 5 ($K_2Cr_2O_7$ series) at 35% solids concentration in turpentine, that the color can scarcely, if at all, be perceived in a film approximately 0.001"-thick.
21. R. M. Johnston and R. L. Feller, "The Use of Differential Spectral Curve Analysis in the Study of Museum Objects," *Dyestuffs* 44, No. 9 (December 1963): 1-10.

7
PROPERTIES OF MATURE VARNISH

PROTECTION BY VARNISH

Picture varnish is generally referred to as a "protective coating;" however, when an attempt is made by technical analysis to determine just what its protection is, we find that there is much that is not known. Five ways will be considered in which a varnish may or may not protect a painting.

1. Protection from Dirt and Abrasion

This is perhaps the most obvious and effective protective action of picture varnish; the varnish prevents dirt from coming into contact with the paint and protects it from abrasion in handling and dusting.

A practical lower limit to the hardness of varnish exists; for example, if the varnish is too "soft," it may defeat its purpose by picking up dirt. Reference here is not to the problem of the electrostatic attraction of dirt, but to the phenomenon that our laboratory first observed with poly(n-butyl methacrylate). At room temperature, poly(n-butyl methacrylate) is above its *second-order-transition temperature*; below this temperature, the polymer is in its glassy state, relatively hard and brittle. At temperatures higher than the second-order-transition point, the polymer is in a state which approaches that of a highly viscous liquid. When films of poly(n-butyl methacrylate) are stored for six months in a particularly dirty area, dirt will become embedded in the film in such a manner that soap and water cannot clean it off; this problem may be encountered with any resinous material that is at temperatures much above the second-order-transition or glass temperature.[1] Poly(vinyl acetate) AYAB and AYAC behave in the same manner; when these have been used as a retouching vehicle, they are usually covered with a final varnish that will not take up dust. Final varnishes should be avoided if their second-order-transition temperature is below about 25°C. (in tropical countries, this minimum should be raised).

2. Protection During Normal Movement of a Painting

A protective coating must be capable of reasonable elongation so that it will remain continuous; if cracks develop, dirt, water vapor, and other agents can enter the crevices. The elongation that a film can undergo before rupture is dependent upon the *rate* at which the stress is applied. However, on the basis of the usual test procedures (strain-rate about 0.2" per minute), we should perhaps consider for picture varnish coatings that are capable of 1 to 3% elongation-at-break.

A varnish must be able to stretch and change its dimensions in such a manner that a minimum of stress is placed upon the paint. In the simplified diagram in Figure 7-1, one may say that the slope of the stress-strain curve

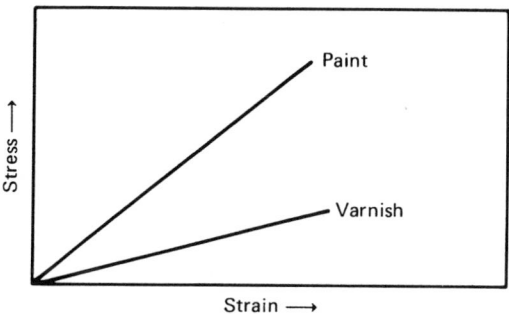

Figure 7-1   Desired Relationship Between Stress Developed in a Varnish vs. That Developed in Paint to Be Protected.

(the elastic modulus in a perfectly elastic material) should be lower for the varnish than for the paint. The problem has not yet been approached in such precise terms, but data in Table 7-1 suggest that experimental varnish 27H possessed properties, chosen on an empirical basis, that were close to this specification. The elastic moduli of a number of paints, reported in the literature, suggest that dried linseed oil paint usually has a modulus much higher than that of the experimental polymer.[2,3,4]

3. Protective Value of a Varnish That Requires "Mild" Solvents and Infrequent Removal

In the long run, one varnish may be more protective than another with respect to possible harm to the painting if it does not require frequent re-

TABLE 7-1. STRESS-STRAIN PROPERTIES OF COATINGS

REF.	COATING	ELONGATION AT BREAK	APPROX. STRESS AT 1% ELONGATION
	Mastic, Dammar	~0.2%	1300 to 3200 psi
	Experimental 27H	1.5	880
2	Paints, 12-weeks old	2 to 5	140 to 800
3	Architectural Paints		1780
4	Alkyd Paints	1 to 4	3400 to 14000

moval; each time a varnish must be removed there is a risk of mechanical and solvent action upon the paint. In addition, if the solvent required to dissolve, formulate, or remove a varnish presents less risk of solvent action upon a painting than does the solvent for another varnish, the first then represents an advantage over the second.[5,6] Thus, because (a) a particular varnish may require milder solvents than another, and (b) it may require less frequent removal than another, the one can be considered to offer greater protection to a painting than the other.

4. Protection from Radiation

Figure 7-2 shows the transmission characteristics of methacrylate and vinyl acetate polymers, well known for their transparency to the shorter wavelengths of radiation. Dammar and mastic varnishes remove more of the ultraviolet radiation than these polymers, and when discolored with age, they remove an even greater portion of the short-wavelength radiation.

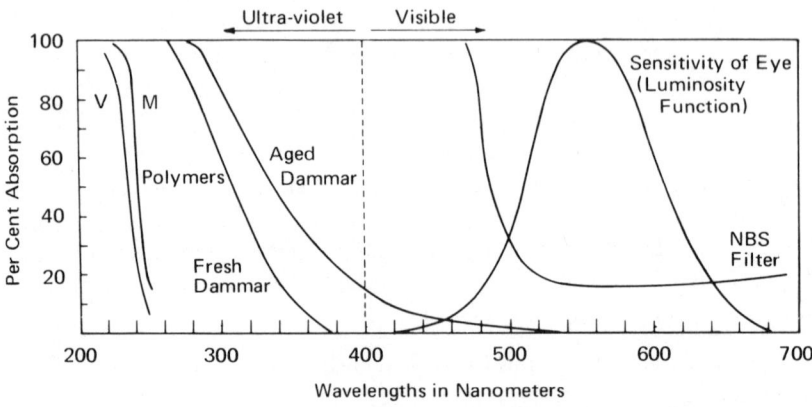

Figure 7-2 Absorption Spectra of Various Materials Compared to Sensitivity of Human Eye.
V = Poly(vinyl acetate) M = Methacrylate polymer, Film Thickness: 0.001"

Undoubtedly, we should not seek to solve the protective varnish problem by placing the picture under a colored filter, as was done with the decidedly yellow filter used to protect the Declaration of Independence and the Constitution of the United States of America ("NBS Filter" in Figure 7-2, U.S. National Bureau of Standards Circular 505, 1951).[7] With documents such as these, legibility and not color, is most important. In the case of paintings, however, we wish to see the true colors. Nevertheless, the eye is not sensitive to radiation less than about 400 nm. and there is no need for the varnish to transmit radiation of shorter wavelength. At present, there are a number of compounds available that are designed to filter ultra-violet radiation. Several of these are suitable for incorporation in picture varnishes formulated with colorless synthetic resins, but as yet they have rarely been applied, one reason being the difficulty, due to the thinness of the usual varnish, of introducing sufficient U.V. absorber to make it an effective filter and yet not run the risk of the absorber "migrating" out of the film.[8] In contrast, plastic sheets which are used as U.V. filters usually can be of sufficient thickness that the low concentration of absorbing compound presents little difficulty with respect to solubility and migration. In Table 7-2 are shown the calculated factors of damage per footcandle (D/fc.) for several coatings and the traditional varnishes; the D/fc. factor has been discussed in the Harrison Report published by the Metropolitan Museum of Art in New York.[9] Note that window-glass filters much, but not all, of the damaging radiation. The negligible "yellow" color of fresh varnish was previously mentioned in Chapter 6; data in Table 7-2 indicate that it is no more protective when fresh than a colorless methacrylate coating (in which the solvent and other components

TABLE 7-2. DAMAGE FACTORS PER FOOTCANDLE OF ZENITH SKYLIGHT AFTER PASSING THROUGH VARIOUS FILTERS*

SOURCE	D/fc
Zenith Skylight (Z.S.L.)	4.80
Z.S.L. through window glass	1.58
Z.S.L. through window glass, varnish 27H	1.45
Z.S.L. through window glass, fresh dammar	1.45
Z.S.L. through window glass, varnish 27H + 0.4% U.V.49**	0.97
Z.S.L. through window glass, aged mastic	0.91
Z.S.L. through Plexiglas ® LPC-518K[9] (now U.F.1.)	0.41
(Daylight fluorescent lamplight for comparison)	(0.40)
(Incandescent lamplight for comparison)	(0.14)
Protected U.S. Declaration of Independence (yellow N.B.S. Filter)	0.02
Limiting factor, removal of all Z.S.L. radiation shorter than 400 nm.	0.46

*Data principally from Harrison.[9]
**American Cyanamid "U.V. Absorber No. 49," 2-hydroxy-4-methoxybenzophenone at 0.4% concentration based on weight of resin, in an experimental varnish designated as 27H.

cause more absorption of the ultra-violet than exhibited by the pure polymers in Figure 7-2).

It is possible to calculate the theoretical damage factor per footcandle of a filter that removes all radiation lower than 400 nm. and transmits all radiation of longer wavelength; that is, a filter, which one might say, is *ideally designed* to remove the wavelengths shorter than 400 nm. and pass all the visible wavelengths. This factor is 0.46 and is given in the table. Thus by research it is possible to develop a colorless varnish which will improve the damage factor per footcandle indoors from the value of 1.45 in present varnishes to somewhere near the theoretical minimum, 0.46. A colorless experimental varnish made in our laboratory (see Table 7-2) provided more protection than fresh dammar and nearly as much as discolored mastic.

5. Considerations Regarding the Permeability of Picture Varnish

Measurements of permeability refer to the *rate* of transmission through a film. It should be emphasized that organic coatings generally do not prevent transmission; they are seldom complete vapor barriers. Instead they reduce the amount transmitted in a given time. Of course, the rate of transmission may be sufficiently low that little change in conditions takes place in a brief period of exposure, such as, in the protection of bread with plastic wrappers. In conservation, however, we are concerned with longer periods of protection and the rate of transmission must be kept in mind.

The consequences of variations in the permeability of picture varnishes to water vapor and other gases have not been demonstrated as far as the author knows. Of course, when stable conditions of temperature and humidity are absent, the effect of reducing the rate of dimensional change in the support has been well-demonstrated; this is particularly advantageous in the case of wooden panels.[10,11,12] However, oil paintings on canvas are not often protected from the rear, unless relined with a wax-resin adhesive. Moreover, when the atmospheric conditions are stable, as in an air-conditioned gallery, it may be expected that water vapor will pass through the usual picture varnish until moisture equilibrium is established in the paint and ground.

The industry engaged in protection of metals once devoted considerably more attention to permeability measurements than it does at present. Investigation soon revealed that more oxygen and water vapor are transmitted by organic coatings than are needed to satisfy the observed rate of oxidation of the metal; that is, more of these reactive substances are able to pass through the film at any time than are actually consumed by the metal.[13] A similar situation may exist in the case of varnishes which protect oil paintings; probably more oxygen and water vapor are transmitted by the traditional varnish than is necessary to account for the slow rate of oxidative destruction of the paint.

The effect of water vapor is not the only concern in the protection of pictures. Gases such as oxygen and hydrogen sulfide also act upon the painting. Stannett and Szwarc have suggested, however, that the relative permeabilities of various coatings to water vapor will reflect their relative permeabilities with respect to other gases.[14] Undoubtedly, protection of the surface of the paint from high concentrations of oxygen and water vapor is one of the significant roles of varnish because we know that photochemical oxidation deteriorates the surface of paint. Unfortunately, at present, there is little experimental evidence available to contribute to a lengthy discussion of this subject.

TABLE 7-3. RELATIVE PERMEABILITY TO WATER
Grams Water/100 in.^{2}/Day at 39.5°C., 100% R.H.
Film Thickness: 0.001"

FILM	GRAMS OF WATER	REFERENCE
Saran, poly(vinylidene chloride)	0.3–1.5	a,e
Paraffin wax	0.5–0.8	a,b,e
Poly(vinyl chloride-acetate)	4–11	a,e
Dammar, Mastic (one-month old)	5–6	b
Polystyrene	25–35	a,b,e
Linseed oil (one- to six-months old)	28–35	b,c
Methyl methacrylate polymer	35	a,d
n-Butyl methacrylate polymer	56	a,d
Isoamyl methacrylate polymer	42	d
Poly(vinyl acetate)	70–180	b,d

(a) P. W. Morgan, *Ind. Eng. Chem.* 45 (1953): 2296–2306.
(b) R. J. Gettens and E. Bigelow, *Technical Studies in the Field of the Fine Arts* 2 (1933-34): 15–25.
(c) A. V. Blom, *J. Oil & Colour Chemists' Assoc.* 22 (1939): 104–19.
(d) Film Properties Fellowship, Mellon Institute (Private communication).
(e) F. Houwink, *Elastomers and Polymers* (New York: Elsevier, 1950), p. 324.

Table 7-3 compares the permeability of a number of materials to water. Under the conditions of this table, Morgan considers films with values of less than 1.5 g./100 in.^{2}/day to possess low permeability.[15] A similar table has been presented by Thomson.[16] Values of permeability vary considerably and are significant primarily in their order of magnitude; Barrer has drawn attention to the fact that a range of 10^5-fold exists between the permeability of water through wax, on one end of the scale, and through textile fabrics or paper on the other.[17]

Materials showing a completely amorphous structure will probably always be found to have rather high moisture permeability.[15] In Table 7-3 the permeability of the methacrylate polymers is among the highest; still, these exhibit lower permeability than many paints and varnishes.[13] There can be no argument that coatings of wax have a low rate of transmitting water vapor.

In view of the following: (a) the lack of evidence regarding the role of permeability in the protection by picture varnish; (b) the variable factors in practice, such as the frequent lack of protection from the rear; and (c) the limitations in the permeability expected in amorphous materials, measurements and comparisons of the permeability of individual picture varnishes cannot be discussed profitably at the present time.

To summarize the protective action of picture varnish: a varnish protects a painting from dirt and abrasion. It also provides protection if it is able to follow the movements of the painting without cracking, at the same time giving rise to a minimum of stress upon the structure of the painting. A varnish may be considered to represent greater safety in use the less frequently it must be removed and the "milder" the solvents that are necessary for its application or removal. It should filter some of the radiation striking the painting and reduce the amount of gaseous substances which reach the surface of the painting in a given time.

SPECIFICATIONS FOR PICTURE VARNISH

Having considered several aspects of protection, we may attempt to specify the properties that are desired in a picture varnish. This objective has its limitation, of course, for no two paintings and protective problems are alike. Nevertheless, at the 1930 Rome Conference, a committee on the restoration of paintings proposed the following specifications:[18]

1. It should be transparent and colorless.
2. Its cohesion and elasticity should be such as to allow for all ordinary changes in atmospheric conditions and temperature.
3. It should be easily removable.
4. The elasticity of the paint film and tissues under the varnish should be preserved.
5. It should protect the painting from atmospheric impurities.
6. It should be capable of being applied thinly.
7. It should not bloom.
8. It should not be glossy.

More recently, Thomson proposed four properties for consideration:[19]

1. The varnish must form a colorless and transparent film.
2. It must be stable over very long periods of time.
3. It must protect the picture from surface deposits and moderate abrasion.
4. It must be possible to apply the varnish with full control.

The order of presentation in both listings has been changed to permit the similarities to be seen more clearly and to align them more closely in order of feasibility. There is little point at this stage in debating the relative importance of the objectives, yet some attempt at listing the requirements is needed by anyone who intends to investigate the subject.

For many years the laboratory at Mellon Institute has had a set of specifications of its own to guide the research, yet attention shifted to the problems of deterioration and stability, and an appropriate occasion to discuss the specification never presented itself. In September 1967, however, a list was presented to colleagues in Italy of five specifications that we had established in the search for new solvent-type coatings for the protection of paintings and for retouching.[20] The added details in brackets were not in the original publication:

1. The resin must be transparent, colorless and resistant to discoloration. [The factor of damage-per-footcandle as calculated by Harrison should be as close to 0.46 as is compatible with the other specifications.]
2. The composition of the resin must be known. In other words, the composition of the materials must not be known only to the manufacturer. This specification resulted in the selection of well-defined classes of polymers for major study: the methacrylate family, poly(vinyl acetate), cellulose acetate-butyrate, poly(vinyl alcohol), and poly(ethylene glycol). The composition of such resins is well known; they are available from a number of manufacturers, and they can be duplicated in the laboratory. The usual practice in America has been for the conservator to make his own solvent-type varnish from polymers and solvents of known composition.
3. The polymer must be flexible in itself, sometimes called "internally plasticized"; the so-called "volatile" plasticizers of the type similar to dibutylphthalate have been avoided. [The dry protective film, nevertheless, must not pick up dirt owing to the fact that it is below its second-order-transition temperature under normal conditions of usage.]
4. The resins must be soluble in relatively "mild" solvents if used on "oil" paintings. Xylene has been set as a standard of the maximum solubility parameter of solvent that would be sought. [Mixtures of methylcyclohexane/xylene and xylene/acetone can be used as a measure of the relative success in meeting this specification.]
5. One final requirement of paramount importance: the coating must remain highly soluble so that it may be removed from the work of art at any time in the future.

The more detailed specifications in brackets have been added because we believe it is possible to be specific regarding these particular points. The last specification cannot be precisely defined at this time if it is admitted that some fractional insolubility in solvents can be tolerated. Polymers that are almost completely insoluble can often be removed easily if they form a soft, highly swollen, gel; it probably is a sound research policy, nevertheless, to avoid considering films that may become more than 25 to 30% insoluble in a mixture of equal volumes of xylene and acetone after 5 million footcandle-hours' exposure at room temperature to diffuse skylight.

Perhaps Thomson's properties 3 and 4 might be added, but the five specifications above are specific and can serve as a point of departure in future studies of the problem.

DETERIORATION OF VARNISH

This subject may be divided into physical deterioration and chemical deterioration.[21] From the viewpoint of modern research, it is usually possible to attribute the physical effects observed, such as, the cracking and chalking of paint, to chemical processes which are occurring within the film. Therefore, the major part of this discussion will concern chemical aspects of deterioration. The possibility must not be overlooked, however, that purely physical changes can be responsible for certain effects noticed in the deterioration of coatings. Changes such as the separation of two physical phases, the syneresis of a gel, or the release of stresses within the film, are physical phenomena that need not be due to chemical changes.

There are one or two pertinent remarks to be made about physical change in a varnish without entering upon a lengthy discourse. The first concerns cracking versus crystallization. When a varnish breaks up into powder, the change has sometimes been called crystallization. Crystallization in its strict physical sense is a phenomenon that seldom occurs unless a substance is in a high state of purity. When films of amorphous materials break up, we usually observe *disintegration*, not crystallization as chemists and physicists use the term. We have called attention to an interesting phenomenon of *interlayer disintegration* of retouches done in dammar varnish mixed with zinc white.[22]

Another important phenomenon in certain varnishes concerns shrinkage similar to that which occurs in mud flats along a river. The author believes that dammar and mastic varnishes in particular, or mixtures of these with drying oils, tend to form a gel. Shrinkage is a well-known property of gels, and we should not overlook its occurrence here. We really know very little about the shrinkage problems in relation to the deterioration of picture varnish. It remains a major area for investigation.

Thomson has discussed at length the phenomenon of wrinkling of varnishes. This seems to be a special problem with resins of low viscosity grade; for example, dammar, mastic, AW-2, MS-2A.[19]

Blooming and blanching of the varnish surface are other common types of deterioration which deserve extensive study. Of bloom alone there are several types.[23] Nevertheless, to the naked eye, inspecting the pictures while passing through a gallery, many of the phenomena of blooming and blanching appear to be superficially similar. Brommelle has reviewed evidence of the role of ammonium sulfate in a widely encountered type of bloom.[24]

What are some of the principal chemical reactions involved in deterioration, reactions that are considered to be responsible for the physical changes observed? Oxidation is the most common. Bear in mind, however, the chemist's definition of oxidation: "a reaction involving the loss of electrons"; oxygen is not solely responsible for reactions in this classification. Ozone, hydrogen peroxide, chlorine, sulfur dioxide, and other gases in the atmosphere can also bring about oxidation.

In the remarks that were made regarding the durability of resins, it was mentioned that some are more prone to oxidation or other chemical reactions than others. For example, mastic seems to oxidize more readily than dammar.[25] Thomson has followed the oxidation of various resins by observing the change in their infra-red spectra.[16]

The occurrence of yellowing in a film is due to certain chemical reactions; very often oxidation is the reaction chiefly responsible. Elm, cited in Chapter 2, has postulated that the formation of phorone-type chemical structures in drying oils is the basis of the yellow color observed as these materials age. In this respect, there is still much to be studied regarding the discoloration of the natural resin varnishes. The role of turpentine, or the residue left by it, in giving rise to discoloration must not be discounted; methacrylate coatings prepared with turpentine often discolor far more than those made with stable and refined-petroleum solvents.

Let us consider specifically the deterioration of linear polymers. Many types have great stability, yet it is important to understand the mode in which even these stable substances deteriorate. We know of at least four ways in which these substances may become altered: 1) by *chain breaking*, that is, the polymeric chains may break up into smaller units. This is the explanation given for the embrittlement of paper, linen, silk, and wool; 2), 3) may be the *general chemical oxidation* and eventual breaking down (degradation) of atomic arrangements within the polymeric substance. *Degradation* may take the form both of the breaking off of low-molecular-weight fragments and the breaking of the main chains; and 4) *cross-linking*, where, in the course of aging, should an atom or group of atoms on a linear polymeric chain become activated, it is possible that the activated groups may link with a neighboring chain. Literally, the linear chains cross-link and become joined one to the other. In effect, this produces molecules of ever-increasing molecular weight, leading eventually to a network of attached molecular chains which are insoluble, a characteristic of tri-dimensional or thermosetting polymers. Such a structure is related to that of linoxyn, aged linseed oil. However, the smaller the molecules of resin to begin with, the less probable will be the buildup to large molecules.

Under sufficiently severe exposures, organic linear-polymers become altered in one or another of the above modes. The total rate of change may be considered as the sum of the rate of chain breaking, plus the rate of cross-linking, plus the rate of oxidation, plus the rate of any other reactions that may be taking place. In certain polymers, or at certain stages of the process, one type of reaction or another may predominate. In early investigations, Dr. L. A. Wall of the U.S. National Bureau of Standards emphasized that the cross-linking reaction tends to predominate in polystyrene and polyethylene. On the other hand, cellulose molecules undergo extensive chain breaking. A "depolymerization" reaction which yields nearly pure monomer is a characteristic of methacrylate polymers heated at elevated temperatures.

Cross-linking is a fundamental and sought-for property of thermosetting resins. The molecules of such substances are intentionally prepared with sites that facilitate cross-linking. While this is desirable in many applications, we would not wish cross-linking to occur to a significant extent in picture varnish. However, with age, certain linear polymers may take this path of deterioration among the alternatives that follow activation of their molecules by heat or light. Because the subject is of particular pertinence, the cross-linking of methacrylate polymers has been studied extensively by the National Gallery of Art Research Project. Although the problem will be summarized in the following paragraphs, details will be found in the appendices.

Appendix A reviews the early history of the detection and analysis of the problem, remarks that appeared in the body of this chapter in the first edition. Appendix B, which appeared in the first edition, discusses the changes in solubility and swelling ability that take place in a film that exclusively cross-links. Appendix C is essentially a reprint of our 1966 publication that discusses the case in which a significant amount of chain breaking as well as cross-linking takes place.

In the summer of 1955, while studying the depolymerization of poly(isoamyl methacrylate) at high temperatures, Mr. Stuart Raynolds of our laboratory noted that the heated samples had become insoluble. The fact that methacrylate polymers might become insoluble under heat and light was little appreciated at the time, but the possibility of this occurrence was a matter of considerable importance in conservation and a program of investigation was soon set in motion. What we managed to learn by the time of the Oberlin conference in April 1957, was presented in the first edition of this publication and announced at a meeting of the American Chemical Society in September.[26] The information presented at that time may be found in Appendix A.

The first studies of the problem involved measurement of the relative ease of removing artificially aged films with cotton swabs dipped in solvent. The results, summarized in Table 7-4, quickly showed that a wide difference existed in the sensitivity of thermoplastic methacrylate polymers toward loss

TABLE 7-4. INCIDENCE OF CROSS-LINKING IN THERMOPLASTIC POLYMERS OF METHACRYLATE ESTERS EXPOSED IN CARBON-ARC FADEOMETER AT ABOUT 60°C.

ALCOHOL RADICAL, R, IN POLY(R-METHACRYLATE) Tertiary hydrogen atoms in boldface	HOURS (a) EXPOSURE BEFORE BECOMING RESISTANT TO REMOVAL IN:	
	CYCLOHEXANE	TOLUENE
50/50 co-polymer of p-methylcyclohexyl and 3-methylbutyl	50	130
3-Methylbutyl	84	160
2-Ethylbutyl Isobutyl	125	200
2-Methylbutyl	148	300
n-Butyl	425	560
n-Propyl		1050

(a) The films were coated on aluminum foil. The time reported is the actual hours of exposure multiplied by 1.9 to allow for the effect of reflection from the aluminum foil.

If a film 2 mils or less in thickness is not dissolved or is not sufficiently swollen to be removed by the friction of a swab after 1¾ minutes' exposure to a certain solvent, it may be said to be resistant to removal in that solvent.

of solubility when exposed in the carbon-arc fadeometer (using Corex-D glass). The investigation also revealed the most sensitive materials to be those polymers based on methacrylic esters which had tertiary hydrogen atoms in the alkyl radical of the alcohol group (the bold H's in Table 7-4).[27] It was found that many of the methacrylate polymers of particular interest to conservators, for example, the isoamyl, isobutyl, and normal butyl, were among those most prone to become insoluble. Thomson reported in 1956 that poly(vinyl acetate) did not cross-link[28] and we suggested in the first edition that other types of polymers would be found that did not have particularly sensitive locations for the development of cross-links; poly(n-propyl methacrylate) was one such polymer (reported in Table 7-4). In 1963, we were able to report that a co-polymer of methyl methacrylate and ethyl acrylate was highly resistant to cross-linking.

Another aspect of the problem that was answered in part by September 1957 was the response to the practical question: what did this phenomenon, discovered and analyzed on the basis of accelerated aging tests, mean in terms of exposure under ordinary conditions? One estimation of the rate of deterioration under normal conditions was made by comparing the rate of cross-linking and the fading of dyes in the fadeometer to the reported rate of fading of the same dyes on a gallery wall (Appendix A); another compared the loss of solubility during outdoor exposure with the estimated relative exposure in a sky-lighted gallery. A third method of estimation was based on a semi-quantitative, paper-chromatographic test developed on fadeometer-exposed samples and then applied to samples that had been at the Fogg Art Museum, Harvard University, for ten years. Although poly(n-butyl methacrylate) had been of interest to museums for no more than about twenty years in 1957-59, and the Fogg Art Museum samples were no more than ten years old when tested (see Table 7-5 for the results of the latest tests on these panels), we estimated in 1957 that the poly(n-butyl methacrylate) and poly(isoamyl methacrylate) would not develop resistance to removal in methyl cyclohexane (or certainly "mild" mixtures of about 50/50 of methylcyclohexane/toluene) for at least twenty-eight years of "ordinary gallery exposure." Moreover, there were tests that could be made to determine the extent of cross-linking before it became serious, and the possibility of inhibiting the reaction with suitable compounds was shown. The problem thus did not appear to be an immediately alarming one, yet the phenomenon clearly required full consideration in conservation research.

In the decade following the discovery of the cross-linking phenomenon and the publication of the first edition of this book, there has been a continuous investigation of various aspects of the deterioration of acrylic polymers by the National Gallery of Art Research Project. The results of this research now

TABLE 7-5. SOLUBILITY OF METHACRYLATE COATINGS TAKEN FROM TEST PANELS ON WALL OF LABORATORY AT FOGG ART MUSEUM

COATING	YEARS EXPOSURE TO SKY-LIGHT ILLUMINATION	PER CENT INSOLUBLE IN 7/3 TOLUENE/ ACETONE	REMOVED EASILY IN THE FOLLOWING TOL/ MCH MIXTURES (a)
Isobutyl, Elvacite ® 2045	12	7.0	1/1
Isoamyl, 27H + 0.4% Uvinul ® 400	12	3.8	1/1
Isoamyl, 27H	14	11.0	1/1
n-Butyl, du Pont Lucite 44	14	33.0	1/1(b)
n-Butyl, du Pont 934	22	53.1	3/1(b)
n-Butyl, polymer 35	22	57.0	1/1(b)
n-Butyl/isobutyl, du Pont 935	22	36.0	1/1(b)

(a) TOL is toluene, MCH is methyl cyclohexane, proportions by volume.
(b) Gel character noticeable during removal.

permit us to describe in considerable detail the manner in which these polymers deteriorate and the factors in their composition that contribute to the greater stability of one polymer over another. As stated earlier, there are at least four ways in which these substances may become altered: in the overall process, which may be regarded as primarily one of oxidation, it is possible to observe a gradual loss of solubility in the original solvents, a loss in weight, and a process of chain breaking and/or cross-linking. The total process can be divided up into perhaps four stages or periods in the history of the film's age. These have been outlined schematically in Appendix Figure B-3. The reader may wish to refer to this figure in the course of the discussion which follows.

In the oxidative deterioration of most organic substances, there is usually first a period of *induction*. This is the period during which those reactions of interest to the investigator may be deferred because there are substances in the protective coating which are attacked more easily by oxygen than the main polymeric material. These substances are oxidized preferentially until their concentration is so low that the slower rate of reaction of oxygen with the polymer becomes significant. An induction period in linseed oil before it begins to oxidize has long been known.

While most materials contain traces of substances that act as inhibitors, substances can also be added intentionally to organic systems to scavenge the oxygen and inhibit reactions which one wishes to minimize. There is a whole field dealing with the subject of antioxidants which should be explored in conservation problems and which undoubtedly can be of assistance in suppressing the tendency of polymers to cross-link. Our work with substituted benzophenones, described in Appendix A, showed that it is possible to inhibit the cross-linking tendencies of poly(isoamyl methacrylate). We have not concentrated on this technique as yet because we considered it wiser to select polymers that are fundamentally stable before attempting to improve their qualities by inhibition. (One reason for this philosophy is that, once the inhibitor has been used up through its preferential reactivity with oxygen, then the reaction that one has sought to suppress will take place.) In seeking stable protective coatings for conservation, it was considered most important to study the properties of the polymers themselves, and to develop types that are particularly stable. After the most stable types have been determined, the problem of inhibition can be taken up as a supplementary investigation.

In the buildup of insoluble matter through cross-linking, there is another reason for an apparent induction period before the appearance of insoluble matter occurs. On the average, one needs about one cross-link for every four molecules before the polymer becomes partially insoluble.[29] It does not take many links before high-molecular-weight materials become insoluble, but it will take many more linking reactions to join together the molecules of a low-molecular-weight resin. The induction time that is controlled by the growth of the molecules to a size sufficiently large to become insoluble, is thus inversely proportional to the average molecular weight of the particular polymer. This is one reason why the very-low-molecular-weight resins, such as dammar, mastic, AW-2, and MS-2A, do not present the conservator with serious problems of cross-linking while, on the other hand, polymers that have viscosity grades ten or more times higher than these often do present difficulties.

Following the induction period caused by the presence of inhibiting substances, either naturally present or intentionally added, there are usually *oxidation reactions* that proceed steadily throughout the rest of the deterioration process. There is a phenomenon of auto-oxidation which can take place in many organic systems, where the rate of oxidation increases in time. We believe, however, that in the particularly stable polymers, such as the methacrylates, the rate of oxidation is controlled primarily by the activation caused by exposure to light. Therefore, in most accelerated aging tests under strong light sources, the *steady oxidation reactions* have been observed to take place at nearly a fixed rate in time, essentially in proportion to the amount of exposure to light.

The oxidation reactions referred to here are generally involved in what may be called chemical *degradation* reactions in which certain radicals in the polymer structure are attacked by oxygen, changed to ketonic, aldehydic and acidic groups; in an advanced stage of this process, low-molecular fragments of the polymer molecules may be lost. These fragments may be carbon monoxide, carbon dioxide, water, and low-molecular-weight organic molecules having perhaps one to five atoms of carbon in them. One way that these rather steady reactions are observed is by observing the weight loss. It has been stated that measurement of the loss of weight of polymers undergoing accelerated aging tests is a useful indication of their inherent stability.[30] Data obtained by the Research Project on the loss in weight of several of the methacrylate polymers is shown in Table 7-6. Undoubtedly most of

TABLE 7-6. WEIGHT LOSS OF METHACRYLATE POLYMERS EXPOSED ON ALUMINUM IN CARBON-ARC FADEOMETER COMPARED WITH TENDENCY TO CROSS-LINK

POLYMER	PER CENT WEIGHT LOSS IN 100 HRS	TIME TO BECOME 50% INSOLUBLE IN TOLUENE
Poly(p-methylcyclohexyl-co-isoamyl methacrylate)	1.6	35 hrs.
Elvacite ® 2044 (n-butyl polymer)	2.0	44
Poly(n-propyl methacrylate)	0.63	667+
Acryloid ® B-72 (MMA/EA co-polymer, Rohm and Haas)	0.22	2000+

these degradation products derive from the gradual erosion of the surface of an organic coating, but if the reactions proceed far enough (perhaps to the extent of 20% loss in weight), shrinking and cracking of the film can also be noticed.

There is another change noted that may be attributed to the steady processes of oxidation. The coating often becomes less soluble in the solvent in which it was originally soluble and requires "stronger" solvents for its removal. This phenomenon was first reported by us in the case of Acryloid ® B-82 and Rhoplex ® AC-33.[6] There is little doubt that the phenomenon takes place because the ketonic and acidic groups that are formed in the process of oxidation make the polymer more polar. As a result, the polymers that are normally soluble in hydrocarbons may require solvents of higher solubility parameter for removal. This relatively minor oxidation of chemical groups within the resin molecule is not limited to synthetic polymers, of course, but is a change that has long been observed in films of the traditional natural resins.

The periods of *induction* and rather *steady oxidation reactions* are common to the deterioration of all organic coatings. A third process in the deterioration of polymeric coatings can be envisioned, however, pictured as the second period or stage of deterioration in the schematic diagram shown in Appendix B. This may be seen, in the case of polymers that undergo exclusive cross-linking, as the period in which the insoluble matter increases to a point of almost complete insolubility. Following the point of initial appearance of insoluble polymeric gel, most films can be removed relatively easily as a mixture of a highly swollen gel and the polymer that still remains soluble (or readily dispersible as a sol) until a high degree of insolubility or a high degree of "cross-link density" is reached. This relative ease of removal probably takes place during a period that ranges up to about five times the length of time (or more correctly, the amount of exposure) that is necessary before the first appearance of insoluble matter (this is usually called the "gel dose" in radiation chemistry). In other words, the removability of acrylic polymers of DP 100 or higher, which is facilitated by a high degree of swelling, may be expected to last for about five times the period of exposure of the gel dose.

During this stage of the changes that occur in a film after the induction period, both chain breaking and cross-linking can take place simultaneously or one or the other of these processes may predominate. If chain breaking takes place exclusively, the tensile strength of the polymer gradually decreases, as it does in cellulose. However, discussion of this process must await another time. We have been considering the problem of removing protective coatings and therefore are especially concerned with the case in which the polymer exclusively cross-links. This is the phenomenon discussed at length in Appendix B. The rate of build-up of insoluble matter follows a regular mathematical law and reaches a stage of about 90% insolubility after about five times the initial gel dose. During this period we have found that polymers such as poly(n-butyl methacrylate) and poly(isoamyl methacrylate) can usually be removed in their original solvents as a mixture of swollen gel and sol, up to the point at which the film has become about 90% insoluble (see Appendix B).

If extensive chain breaking also takes place at the same time as cross-linking, the rate of buildup of insoluble matter will not be as rapid. Charlesby's equations, mentioned in Appendix C, allow one to estimate the ratio of chain breaking to chain-linking; this is done by measuring the rate at which the soluble matter decreases in a polymer as a function of the time required before the initial appearance of insoluble matter (the gel dose). The wavelength of light influences the relative importance of the chain breaking and chain-linking reactions, and undoubtedly this influence will vary with different polymers. If, as we have shown,[6] the very short-wavelength radiations enhance the cross-linking reaction, it can be seen that the predictions based on the accelerated aging tests may differ from the results found under normal conditions of illumination in galleries. We have also found that high temperatures usually speed up the cross-linking reaction, making it likely that this reaction will be more prominent in the usual accelerated aging devices than in normal gallery conditions (see Appendix C).

As a third stage in the aging of a polymeric coating that exclusively cross-links, the diagram in Appendix B depicts a period in which the film is almost completely insoluble but is able to swell with solvents. This *swelling stage* is governed by the cross-link density (that is, the average molecular weight or degree of polymerization between cross-links) and by the solubility parameter of the solvent causing the swelling.

From the opposite point of view, we may say that the swelling phenomenon is influenced by the solubility parameter of the film (in relation to that of the solvent). This viewpoint helps us to understand that, as the polymer becomes slightly oxidized due to the process of steady oxidation discussed above, it will require "stronger" solvents to cause an equal amount of swelling. When the ability to remove cross-linked films was first studied in our laboratory by measuring the ease of removal of the films with cotton swabs dipped in various solvents, it was pointed out that the removal of the films became progressively more difficult in a solvent such as cyclohexane, then progressively more difficult in toluene, and finally, progressively more difficult in acetone (Table B-1, Figures B-2 and B-3).

The ability of any polymeric network to swell is, as we have said, controlled by the cross-link density. Another way to look at this is to say that it is controlled by the number of monomeric units between cross-links. Obviously, a network which has a long segment of chain between each link is much more flexible and would have much more opportunity to expand and swell than one with much smaller dimensions. As pointed out in Appendix B, the theory that explains the ability of polymer networks to swell is developed extensively. Perhaps we can say that when the number of monomeric units between cross-links falls below about 50, then the usual poly(n-butyl methacrylate) polymer will be extremely difficult to remove.

If the polymer undergoes one chain break for every chain that is linked, after about five times the initial gel dose, it may be only 50% insoluble. Nevertheless, the density of cross-links in the insoluble portion will steadily increase and therefore the insoluble portion will become less able to swell and hence more difficult to remove. As far as we can tell from such information as given in Table B-1, the usual polymers of viscosity grade around twenty to thirty will become very difficult to remove in solvents after about ten times the initial gel dose.

Beyond the stage in which films which extensively cross-link build up an increasingly higher cross-link density, thereby losing their ability to be swollen (which we have shown to be one of the controlling factors in removability rather than mere insolubility [Appendix B]), we must envision a period in which the films cannot be effectively swollen and therefore will become extremely difficult to remove, unless by chance or design they are brittle. With many of the tougher high polymers, the latter is not likely to be the case. Since we know very little about this state of polymeric films (although it is closely related to the state of thoroughly aged drying oils and resin-oil varnishes) we must consider that such films will pose difficult problems in removal. Films in this state, of course, continue to undergo the steady oxidation reactions described above. However, most of these probably take place on the upper surface and therefore the degradation reactions offer little hope that the films will once again become easy to remove until very advanced stages of deterioration are reached.

So much for a general outline of the process by which a thermoplastic protective coating changes upon aging. What does this mean to the conservation problem? First of all, we have said that the cross-linking of the butyl-methacrylate polymers is a phenomenon that can take place under ordinary gallery conditions. Depending upon the intensity of illumination, we have estimated and shown that the films will become perhaps about 50% insoluble in toluene in about twenty-five years under "normal gallery conditions" and in that state will still be readily removed in solvents that are "milder" than toluene. Because ultra-violet light accelerates these reactions, it is best to avoid the use of unfiltered fluorescent lamplight or natural skylight when illuminating objects that have been covered with such coatings. The phenomenon obviously is important enough that research has been directed to finding polymers which do not exhibit this tendency. Nevertheless, where certain polymers seem particularly appropriate to a conservation problem and where they will not be exposed to excessive ultra-violet radiation, we feel that polymers such as poly(n-butyl methacrylate) may still be used in the conservation of many objects, taking into due regard our current extensive knowledge of the phenomena. Tests are also available now, Thomson's very simple test for insolubility and our own technique of measuring the insolubility of material taken up on cotton swabs, which will allow one to monitor the course of these changes in coatings with considerable assurance. Since there is no outward way of noting that the polymers are undergoing a cross-linking change, it is necessary to give particular consideration to the intended application of polymers that may cross-link and to have a fixed program of monitoring the changes in solubility that may take place. Certainly a new polymer that is considered for conservation must be studied in the light of its tendency to cross-link. As stated in 1963, however, the tendency of thermoplastic polymers to cross-link has been found to be more the general rule than the lack of this tendency. It is partly for this reason that such a thorough study of the phenomena was made and particular attention given to the ability of the insoluble portion to swell.

REFERENCES AND NOTES

1. The two temperatures are nearly the same. A polymer seldom has a sharp melting point, but a gradual transition can usually be detected in most, during which the material may be regarded as passing between the state of a highly viscous liquid and that of a brittle glass.
2. W. J. Snoddon and L. L. Carrick, "Cohesive Properties of Soya-Alkyd Unpigmented and Pigmented Protective Coating Films," *Official Digest, Federation of Paint and Varnish Production Clubs* No. 350 (1954): 195-236.
3. S. Werthan, "Trade Sales Paints Formulation and Evaluation," *Official Digest, Federation of Paint and Varnish Production Clubs* No. 374 (1956): 186-205.
4. A. C. Elm, "The Stress-strain Properties of Clear and Pigmented Films of Pure Drying Oil Compounds," *Official Digest, Federation of Paint and Varnish Production Clubs* No. 322 (1951): 701-23.
5. The National Gallery of Art Research Project has been using mixtures of methyl cyclohexane/xylene and xylene/acetone as a convenient measure of the relative "strength" or solubility parameter needed to remove coatings. We have previously suggested that varnishes be sought that are soluble in solvents that possess no greater solubility parameter or swelling action towards linseed oil paints than xylene.[6] This was an attempt to establish a practical working standard of the maximum potential activity of the solvent that should be considered. Tests suggest that xylene alone or a mixture of 25/75 acetone/xylene (by volume) will not remove thoroughly aged dammar varnish.
6. R. L. Feller, "New Solvent-type Varnishes," in *Recent Advances in Conservation* (London: Butterworths, 1963): 171-75.
7. G. Thomson, "A New Look at Colour Rendering, Level of Illumination, and Protection from Ultraviolet Radiation in Museum Lighting," *Studies in Conservation* 6 (1961): 49-70.
8. H. Gysling and H. J. Heller, "Ultra-violet Absorbers as Stabilizers for Plastics," *Künstoffe* 51 (1961): 13-20; *Technica* (Geigy Industrial Chemicals) 3 (December 1961): 5-16.
9. L. S. Harrison, "An Investigation of the Damage Hazard in Spectral Energy," *Illuminating Eng.* 49 (1954): 253-57.
10. R. D. Buck, "The Use of Moisture Barriers on Panel Paintings," *Studies in Conservation* 6 (1961): 9-19.
11. A. E. A. Werner, "Plastics Aid in Conservation of Old Paintings," *British Plastics* 25 (1952): 363-66.
12. A. E. A. Werner, "The Scientific Examination of Paintings," *Lectures, Monographs, and Reports No. 4* (London: Royal Institute of Chemistry, 1952).

13. J. E. O. Mayne, *Research* 5 (London, 1952): 278-83. Investigations have demonstrated that it is not so much the reduced permeability of a coating which retards the corrosion of metals as it is the increased electrical resistance given to the surface. In like manner, the importance or role of permeability in picture varnish cannot be discussed without experimental evidence.
14. V. Stannett and M. Szwarc, "The Permeability of Polymer Films to Gases—A Simple Relationship," *J. Polymer Sci.* 16 (1955): 89-91.
15. P. W. Morgan, "Structure and Moisture Permeabilities of Film Forming Polymers," *Ind. Eng. Chem.* 45 (1953): 2296-2306.
16. G. Thomson, "New Picture Varnishes," in *Recent Advances in Conservation* (London: Butterworths, 1963), pp. 176-84.
17. R. M. Barrer, *Diffusion In and Through Solids* (Cambridge: At the University Press, 1951).
18. R. J. Gettens, "Chemical Problems in the Fine Arts," *J. Chem. Ed.* 11 (1934): 587-95.
19. G. Thomson, "Some Picture Varnishes," *Studies in Conservation* 3 (1957-8): 64-79.
20. R. L. Feller, "Research on Durable Thermoplastic Polymers for the Conservation of Works of Art," in *Atti della XLIX Riunione SIPS, Siena, 23-27 September, 1967* (Rome, 1968): 1099-1110.
21. For a discussion of the mechanical alteration of the paint film, see S. Keck, "Mechanical Alteration of the Paint Film," *Studies in Conservation* 14 (1969): 9-29.
22. R. L. Feller, "Problems of Retouching: Chalking of Intermediate Layers," *Bulletin of the American Group-IIC* 7, No. 1 (1966): 32-34.
23. M. Hess, "Hazing and Blooming," *Paint Film Defects* (New York: Reinhold, 1951), p. 443.
24. N. Brommelle, "Bloom in Varnished Paintings," *Museums J.* 55 (1956): 263-66.
25. R. L. Feller, "The Deteriorating Effects of Light on Museum Objects; Principles of Photochemistry, The Effect on Varnishes and Paint Vehicles and on Paper," *Museum News, Technical Suppl.* No. 3 (June, 1964).
26. R. L. Feller, Paper No. 73, "Cross-linking of Methacrylate Polymers by Ultra-violet Radiation," *Papers Presented at the New York Meeting, Division of Paint, Plastics, and Printing Ink Chemistry, American Chemical Society* 17, No. 2 (September 1957): 465-70.
27. If this theory is correct, the n-butyl methacrylate polymer perhaps should be more stable than it seems to be, but gas chromatographic studies show that the monomer is not completely pure. Impurities that contain tertiary hydrogen atoms, either in the initial monomer, or as oxidized fragments, undoubtedly exist.

28. G. Thomson, "Test for Cross-linking of Linear Polymers," *Nature* 178 (1956): 807.
29. P. J. Flory, "The Structure of Vinyl Polymers; Cross-linking in Diene Polymers," *Principles of Polymer Chemistry* (Ithaca, New York: Cornell University Press, 1953), p. 261, Chapter VI.
30. C. J. Berg, W. R. Jarosz and G. F. Salathe, "Performance of Polymers in Pigmented Systems," *J. Paint Technology* 39 (1967): 436-53.

Part IV

THE REMOVAL OF VARNISH

Elizabeth H. Jones

8
INVESTIGATIONS ON THE REMOVAL OF AGED VARNISH COATINGS

This paper will describe six series of tests, the common aim of which was to determine the ease of removing thirty-one different aged surface coatings, both natural and synthetic, in five different solvents. These coatings had been painted out on prepared panels in 1938. Small areas of these aged coatings were tested with solvents in 1940, 1948, 1961 and 1969. Other areas of each specimen were tested in 1949 and 1961 after previous treatment called "re-forming." The re-forming technique and the results of investigations regarding its mechanism will be discussed in the later parts of the paper.

TESTS ON THE EASE OF REMOVAL OF VARIOUS VARNISHES

Description of Specimens

The Fogg Museum has four sets of specimens of thirty-one different surface coatings which were painted out on test panels in 1938. Of these, twenty-five were fully identified. The value of the aged and identified coatings is greatly increased because a number of synthetic coatings were included. Some of the latter appear to be the oldest identified specimens now available. These specimens are:

1. Mastic
   (a) Spirit varnish prepared in Fogg Museum laboratory (50 g. 1938 mastic, 100 ml. redistilled turpentine, 5 g. polymerized linseed oil)
   (b) Devoe's picture mastic varnish
   (c) Schmincke's #2017 mastic varnish
   (d) Weber's mastic varnish, picture strength

(e) Winsor & Newton picture mastic varnish
(f) Sargent picture mastic varnish
2. Dammar
    (a) Spirit varnish prepared in Fogg Museum laboratory (100 g. Penang dammar, 90 ml. V.M.P. naphtha, 90 ml. redistilled turpentine, 10 g. stand oil in 10 ml. toluene and 20 ml. methanol)
    (b) Devoe's white dammar
    (c) Weber's dammar
3. Copal
    (a) Devoe's picture copal varnish
    (b) Schmincke's #2047 copal varnish
    (c) Weber's copal varnish picture strength
    (d) Winsor & Newton picture copal varnish
    (e) Sargent picture copal varnish
4. Vinyl Resins: solutions of these resins were formulated in the Fogg Museum laboratory
    (a) Vinylite A poly(vinyl acetate). Carbide and Carbon Chemical Corporation, 20% solution
    (b) Vinylite VYHH poly(vinyl chloride-acetate). Carbide and Carbon Chemical Corporation, 15% solution
    (c) Alvar poly(vinyl acetal). Shawinigan Products Corporation, 20% and 30% solutions
5. Methacrylate Resins: solutions of these resins were formulated in the Fogg Museum laboratory
    (a) Poly(n-butyl methacrylate). E. I. du Pont de Nemours and Company, 20% solution
    (b) Poly(n-propyl methacrylate). E. I. du Pont de Nemours and Company, 20% solution
6. Polystyrene (Dow A-25, B-69), Dow Chemical Company, 15% and 25% solutions formulated in the Fogg Museum laboratory
7. Hercose C (cellulose acetobutyrate), Hercules Powder Company, 10% and 20% solutions formulated in the Fogg Museum laboratory
8. Chlorinated rubber (125 centipoises), E. I. du Pont de Nemours and Company, 20% solution prepared in the Fogg Museum laboratory
9. Six proprietary coatings not identified by the manufacturers.

Since groups 6, 7, and 8 are of little interest as protective coatings on paintings, they and the six unidentified proprietary coatings will not be discussed in this paper.

Testing Procedures

One set of these panels was exposed under glass on the roof of the Fogg Museum from July until October 1939, for a total of thirteen weeks. A

TABLE 8-1. THE EFFECT OF AGE ON THE APPEARANCE OF COATINGS.

RESIN	APPEARANCE IN 1940		APPEARANCE IN 1957		APPEARANCE IN 1969	
	PROTECTED BY MASK	EXPOSED	CONTROL	EXPOSED	CONTROL	EXPOSED
Mastics	Yellowed appreciably.	Yellowed like masked section. Hard, glossy, clean.	Markedly yellow. Crazed.	Slightly less yellow than control.	Markedly yellow. Crazed.	Slightly more yellow than control. Crazed.
Dammars	Yellowed appreciably.	Yellowed like masked section. Fairly glossy. Fine crackle net in area 4 coats thick.	Both yellowed equally. Both somewhat crazed.		Markedly yellow. Crazed.	Slightly more yellow than control. Crazed.
Copals	Yellowed and darkened.	Bleached, glossy, hardened.	Yellow to yellow-brown. Several specimens badly crazed.	Slightly less yellow than control.	Both yellow to yellow brown, several specimens badly crazed. Two specimens have collected considerable dust.	
Vinylite A	Not recorded		Faintly grey with dust. Colorless when dust is removed. Slight gloss in both.		Unchanged except for greater accumulation of dust, colorless when dust is removed. Slight gloss.	
Vinylite VYHH	Unchanged	Darkened	Grimy. Colorless when dust was removed.		Grimy. Colorless when dust was removed.	Slightly reddish hue. Grimy.
Alvar	Unchanged	Yellowed	Slightly yellowed	Markedly yellow	Slightly yellowed	Markedly yellow
Methacrylates	Darkened slightly	Unchanged	Grimy. Colorless when dust is removed. Fair gloss.		Grimy. Colorless when dust is removed. Fair gloss.	

section of each panel was covered by an opaque mask. Another set, used as a control, was exposed on the laboratory wall to strong daylight (window and skylight). Since 1947 both sets have been exposed side by side on the laboratory wall. In 1940, R. J. Gettens recorded his observations on the condition of the exposed and the masked areas of the set exposed on the roof. A brief summary of these observations and notes made by the present author are tabulated in Table 8-1.

The third and fourth sets of these panels were exposed on the roof at the same time (July to October 1939) and have been stored in the dark since.

Solvents used in testing: The coatings on these panels have been tested and graded on six occasions (in 1940, 1948, 1949, twice in 1961, and again in 1969) for their relative rates of solubility[1] in five different solvents. The solvents used were: varnish maker's and painter's naphtha (VM&P naphtha), toluene ethylene dichloride, methanol and acetone. Since naphtha had little, if any, effect on the varnishes (except on the methacrylates as noted, page 178), the results obtained with this solvent have been omitted from the graphs.

Methods of Testing: In 1940, the testing procedure was designed to duplicate the normal process of removing a surface coating from a painting. Swabs of standard size were dipped in solvent and rolled back and forth across a 3/4" strip of the coating. The effectiveness of the solvent was graded according to the number of rollings, twenty, forty, or sixty, required to remove a certain fairly constant amount of film. A notation was made if the solvent had no discernible effect.

In 1948, a different method of testing was chosen to study: 1) the time interval required for these same solvents to affect the coatings without the help of mechanical action; and 2) the duration of exposure necessary to swell or to dissolve the coating enough to remove it. The solvent was dropped onto the coating with a syringe. The swelling action was observed under a binocular microscope and timed with a stopwatch. A needle was used to test the degree of softening. Finally, a small amount of the coating was removed with a swab. The gradings given were:

   X—no discernible effect after six minutes
   A—coating affected and removable in six minutes
   B—coating affected and removable in two minutes
   C—coating affected and removable in forty seconds
   D—coating affected and removable in ten seconds or less

The 1948 testing method was used for all subsequent tests. In 1949 and 1961, the tests were repeated seven days after a section of each coating had been "re-formed," a technique described in detail below. The technique consisted of spraying the coating briefly with a mixture of solvents.

Comparison of Grading Systems: The 1940 gradings cannot be compared directly with the gradings of subsequent tests because the testing methods were different, but a roughly equivalent grading system was devised. The translation of the 1940 grades is included on the graphs in Figure 8-1. The changes over the course of twenty-nine years in the effectiveness of the various solvents in attacking the coatings are of interest (Figure 8-1).

Summary of Results

The Natural Resins: Of the coatings based on natural resins, the samples of mastic exhibited the most consistent and marked decline in the rate of attack by solvents between 1940 and 1948. After aging for ten years all were insoluble in toluene and only slowly soluble in ethylene dichloride. Most of the samples remained fairly soluble in methanol and all were quickly soluble in acetone. After aging for twenty-one years, only acetone was effective in removing the coatings as thin sols in less than a minute. The tests made a week after re-forming, in 1949 and 1961, indicated for all samples a remarkable increase in the rate of attack by toluene and ethylene dichloride. There was less chance for improvement with methanol and acetone.[2]

The dammar varnishes, less readily attacked by solvents in 1940, changed less after further aging for eight years, particularly in respect to ethylene dichloride. But by 1961 there was a marked decline in their ease of solubility. Only acetone was effective in removing the coatings without leaving an insoluble residue. After re-forming in 1949 and 1961, toluene and ethylene dichloride increased in effectivness, as they had with mastic. There was a slight increase in the rate of attack by methanol and acetone on the two proprietary samples of dammar.[2] Curiously, methanol had increased effectiveness on both dammar and mastic in 1969.

The copals, which are thermosetting resins, manufactured into oil-resin varnishes, were of limited solubility even in 1940. By 1948, no example actually dissolved or swelled enough to be completely removable in any solvent. In 1961 and 1969, even after exposure for six minutes, they still were tough dry gels, requiring considerable mechanical action to dislodge even a fraction of these coatings. Re-forming had less effect in restoring the original solubility characteristic of these, as of other specimens containing oil.

The Synthetic Resins: Alvar poly(vinyl acetyl) and Vinylite VYHH (vinyl-acetate-vinylchloride co-polymer) were much less satisfactory in their response to solvents than was Vinylite A poly(vinyl acetate). Vinylite A was quickly removed by toluene, methyl alcohol, ethylene dichloride, and acetone in all six series of tests. Naphtha, in which it was not soluble originally, had no effect on it later.

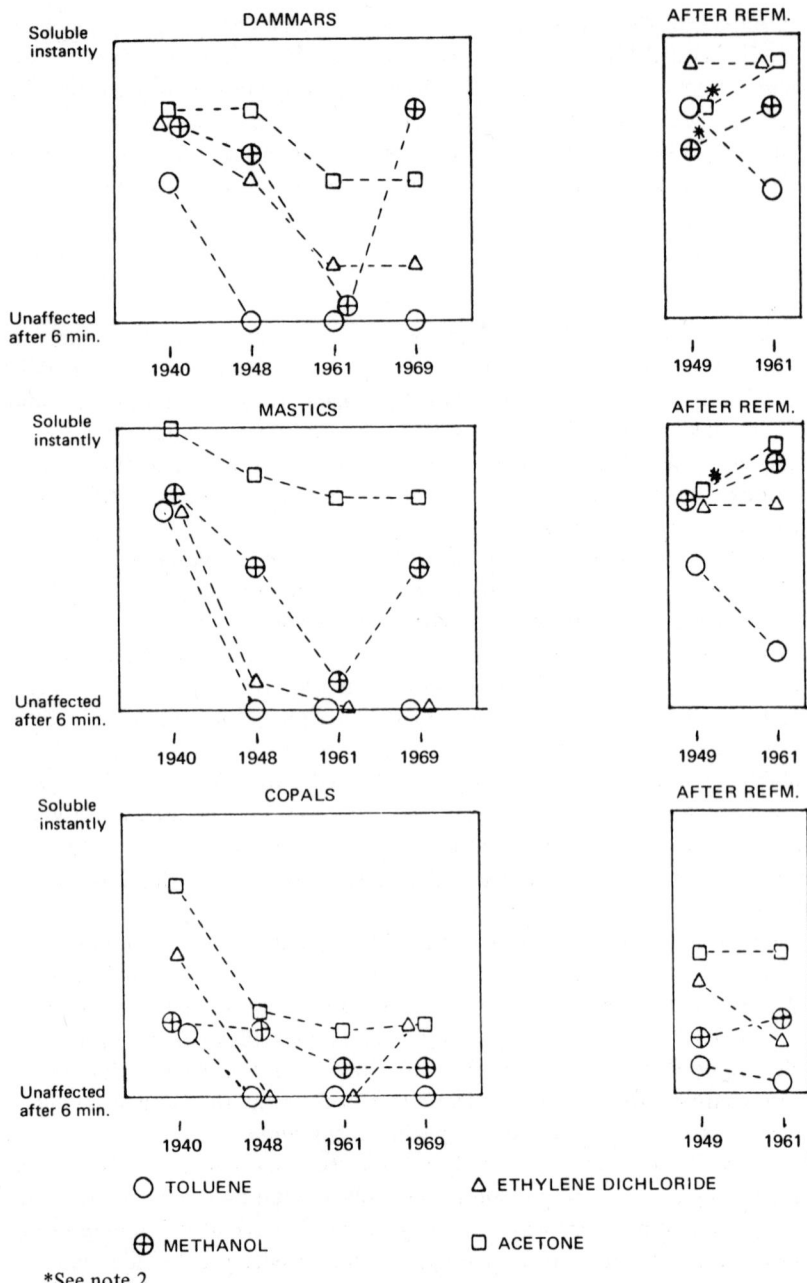

*See note 2.

Figure 8-1 Changes in Solubility of Coatings in the Course of 29 Years.

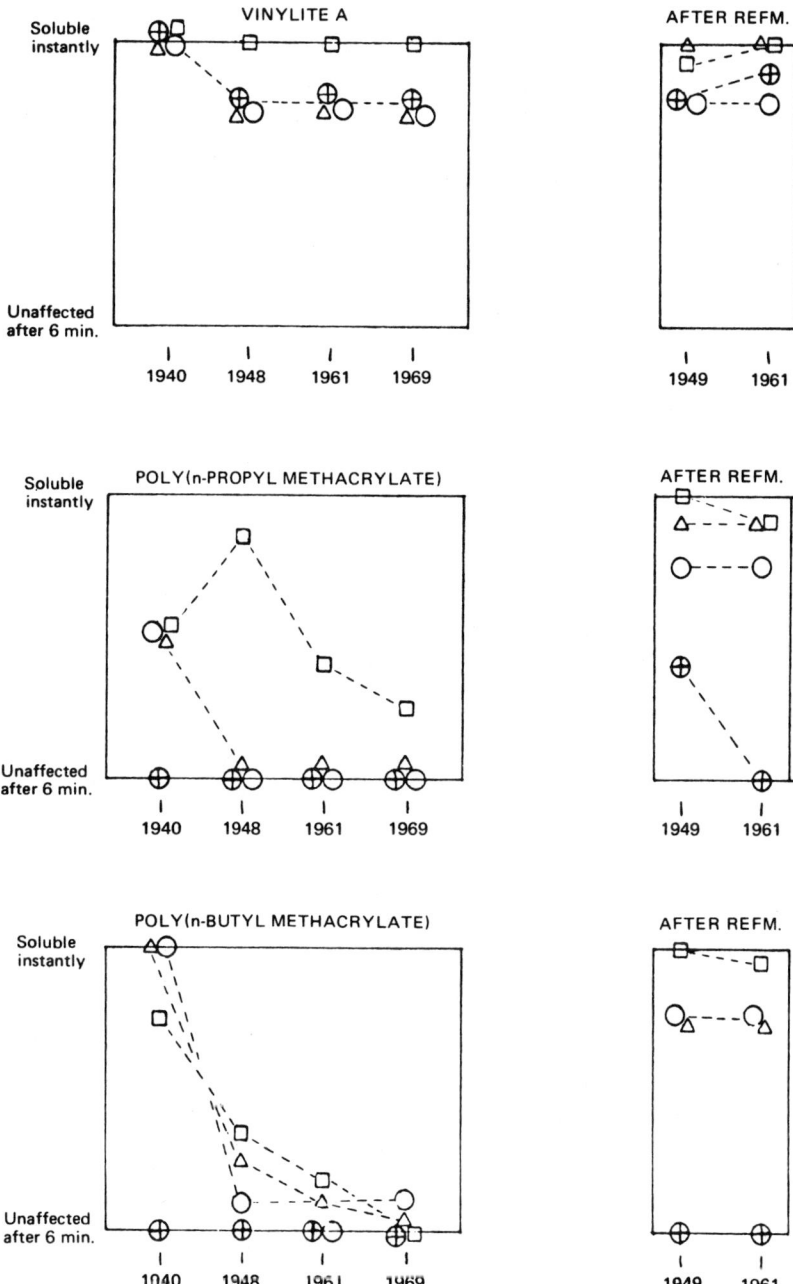

Figure 8-1—Continued

The two methacrylates (n-propyl and n-butyl) which had been exposed to sunlight on the roof between July and October in 1939 exhibited a curious behavior in 1948. At that time they imbibed quantities of solvent, becoming more or less soft swollen gels, but only in one case (acetone on n-propyl) could the coating be easily and quickly removed. This was our first indication of their unusual behavior on aging. Over the years these samples have displayed ever-increasing resistance to solution. Studies by Thomson[3] and Feller[4] have established that cross-linking between molecules induced by exposure to light and heat is the cause of this phenomenon.

A week after re-forming, in 1949 and in 1961, both resins were readily removed in from ten to forty seconds with naphtha, toluene, ethylene dichloride, and acetone. Methanol still had little effect, but these resins were not originally soluble in methanol.[5]

We find, however, that methacrylate coatings applied between 1945 and 1948 to paintings which have been subjected to less intense light in the galleries of the museum and in Harvard lecture halls and common rooms, exhibit another troublesome characteristic. Naphtha, used in the attempt to remove superficial dust and grime, will so swell and disturb these methacrylate coatings that they have to be completely removed with toluene. We conclude that these coatings, exposed to a more normal level of light, have not cross-linked to the degree of the specimens exposed on the roof.

THE RE-FORMING OF VARNISH COATINGS

The Technique

In the preceding section, we have shown it to be possible to increase substantially the rate of attack by solvents on a number of coatings of spirit varnishes by using the "re-forming" technique, that is, by spraying the coating briefly with a mixture of solvents. In 1949, the usual mixture was used, consisting of: ethanol, 4 parts; diacetone alcohol, 1 part; cellosolve acetate, 1 part. In 1961, n-propanol was substituted for ethanol. The spraying is continued only until the coating becomes slightly tacky to the touch. In most cases, this is a matter of exposure for two to four seconds for any given area. Normally the amount of solvent applied by spraying is less than 1% of the amount that would be required to remove the coating if the solvent were applied directly with a swab. After a period of from one hour to one week, removal of the varnish by the usual process is begun. The susceptibility of the paint film beneath to attack by solvents does not appear to be affected by the preliminary solvent spray treatment.

Precedents and History

There are precedents for this technique. In 1864, M. von Pettenkofer took out a patent on his process for restoring the optical properties of aged varnish films by exposing the surface of the picture to alcohol vapor in a chamber.[6] His process had limitations which are now well known: the long exposure necessary, with its risk of swelling the paint film beneath, and the rapid return of the undesirable surface characteristics. In his procedure, no effort was made to remove the grime. The discoloration of the aged coating was probably increased by the application of copaiva balsam, used in an attempt to prolong the effectiveness of the treatment. Those of his contemporaries who used his technique[7] noted that the varnish could be removed more easily after this treatment, if cleaning proved desirable. An improved version of his method has been commonly used in this country and abroad when some immediate improvement in the appearance of the surface coating is desired but the time and expense of the removal of the surface coating cannot be undertaken. In this process of "regenerating" the varnish film, the grime is first removed and the surface is sprayed with solvent, thus avoiding the prolonged exposure in the vapor chamber. Some restorers, on occasion, varnish a painting midway in the cleaning process. The new varnish is permitted to dry, and then removed with the remainder of the old varnish. This technique has been used with some success in making the residues of old varnish easier to remove.[8]

Morton C. Bradley, Jr., first noticed in 1949 that a regenerated coating was more readily removed than it had been before treatment.[9] He suggested that the author test the effectiveness of this preliminary spraying treatment on the same panels, with the same solvents and by the same testing method that had been used in the 1948 tests (that is, by flooding a small area with a syringe and timing the interval until the varnish was visibly affected, and until it swelled or dispersed enough to be removed).

Tests of Re-forming on 1938 Fogg Test Panels

All but a one-inch band across the panels was masked with waxed paper. This area was sprayed briefly, until the resin was tacky to the touch, with the re-forming mixture given above. After seven days, the coatings were tested by the method used in 1948. Definite increases over the tests made before re-forming were noted in the rate of attack by solvents. The changes noted have been discussed previously and are indicated in Figure 8-1.

Summary of Results

The results obtained in the tests after re-forming may be summarized: the re-forming technique appeared to be markedly effective in improving the rate of attack upon most of the coatings by the solvents of low polarity: toluene and ethylene dichloride. The gradings given in 1940 should not be compared directly to the subsequent gradings since a different testing method was used, but it would appear that, on an average, re-forming restored the rate of attack upon most of the aged films more than half way to the rate after aging for a year in 1940. The degree of improvement was less noticeable with acetone (which had retained more of its original effectiveness) and with methanol. Naphtha, as already noted, only very rarely had any effect on these resins at any time.

Practical Application of the Re-forming Technique

Certain conclusions have been reached about the practical application of this technique, first from experiments in the Fogg Museum laboratory on worthless paintings, and later from applications of the technique to the removal of varnish from several thousand paintings. These conclusions may be summarized:

If re-forming is to be effective, the resinous coating should be cleaned of dust, oily grime, or wax.

The original formula for the re-forming spray (ethanol, 4 parts; diacetone alcohol, 1 part; cellosolve acetate, 1 part) still appears to be the safest and most effective. The substitution of n-propanol for ethanol in the re-forming spray used in 1961 produced only minor variations in the results, too slight to be conclusive. Methyl ethyl ketone was also tried as a substitute for ethanol but seemed no more effective. Ethanol, acetone, or diacetone alcohol used alone are not effective.

Exposure of a varnished surface to the vapor of the re-forming solvent and of other solvents in a vapor chamber seems to be less effective (and is certainly much slower in its effect) than the direct application by spraying. Moreover, spraying permits observation and control of solvent exposure, which is difficult or impossible when using a vapor chamber.

In the application of this technique in cleaning pictures, we find that the paint beneath the coating, whether it is tempera, oil, or an oil-resin mixture, does not appear to be affected in any way by the re-forming solvent. The number of effective solvent molecules applied to the coating is limited, far less than the number necessary to disperse the coating in the solvent. A considerable amount of solvent is lost immediately by evaporation from the surface (see Figure 8-3, below). The number of molecules which succeed in penetrating the coating to the paint surface must be very small, far too few to be effective in swelling the paint beneath.

A re-formed resinous coating can be removed with a milder solvent than was necessary before the re-forming treatment, or it can be removed more quickly and with a smaller amount of strong solvent than was possible prior to spraying. With the same solvent, less mechanical action is needed to remove the coating after re-forming, since a less viscous solution is produced; hence, the danger of abrasion is reduced.

On many occasions it has proved possible to clean a painting by repeated treatments (that is, by alternating re-forming and cleaning) with a solvent which was not effective in removing the coating before re-forming. Usually, however, such behavior is not predictable. In a few cases it was clear that there were two or more distinct layers of coating materials applied at different times. Much more commonly, such coatings appear to be one homogeneous layer. It is possible that the top surface of a layer of varnish exposed to light and air discolors first. Then it acts as a filter for the shorter, more energetic wavelengths at the blue and ultra-violet end of the spectrum, shielding the rest of the film beneath to a certain extent.[10] It would seem that the re-forming solvent has penetrated only to a certain depth in a quantity great enough to effect a change in solubility. When the affected layer is removed, the unaffected layer can be reached by another spray of re-forming solvent. If it is necessary at the final stage of removal to resort to a different solvent, such as acetone, or to a very active solvent, such as morpholine, this solvent can be used in more limited amounts than would have been necessary originally, and its effect can be observed and controlled more closely.

A very resistant coating can often be re-formed more effectively if the spraying of the re-forming solvent is followed immediately by a spray coating of varnish. Poly(n-butyl methacrylate) is often used for this purpose. Since it is insoluble in alcohol, it helps to slow down the evaporation of the re-forming solvent. Arthur K. Doolittle, formerly Assistant Director of Research for Carbide and Carbon Chemical Corporation and an authority on solvent action,[11] first suggested this technique.

INVESTIGATIONS OF THE MECHANISM OF RE-FORMING AN AGED VARNISH

Since the practical safety and the effectiveness of this method seemed evident, it became desirable to obtain some understanding of the reasons for the improvement in the rate of attack by solvents on resinous coatings following re-forming. Was it the result of a chemical or a physical change or of the presence of retained solvent?

Infra-red Spectroscopy

Mr. Arthur Doolittle suggested that infra-red spectroscopy might be useful in indicating the nature of the change caused by re-forming and for evidence of the evaporation rate of the re-forming solvent from the resin.

Each organic substance absorbs infra-red radiation at characteristic wavelengths. This absorption can be measured by an infra-red spectrophotometer and recorded in graphical form. The pattern of the graph will be different for every organic substance; these patterns have been called "the fingerprint of the molecule." In a mixture of two organic substances, their patterns will be superimposed, the percentage of absorption at the characteristic frequencies depending on the amount of material present. Any chemical change in the substance will result in an alteration of the absorption pattern.

To help us establish what happens in the re-forming process, a series of infra-red absorption measurements was carried out on two resins. Dammar was chosen as an example of a natural resin. As a synthetic resin, poly(n-butyl methacrylate) was selected since it exhibited striking changes in solubility upon aging and after re-forming. Dr. David Robinson, formerly of Baird-Atomic, Inc., of Cambridge, Massachusetts, manufacturers of a precise infra-red spectrophotometer,[12] directed these tests.

Test Procedure: Two samples of each of these resins, 20% solutions in toluene, were brushed out on sodium chloride crystals (which do not absorb light in the infra-red range of wavelengths). At the same time, these same solutions were painted out on test panels for parallel studies on their solubility.

Immediately after coating, infra-red absorption patterns were taken of the resin coatings on the salt crystals. In addition to their characteristic absorption patterns, there was a strong absorption band at 13.7 μ, an absorption band characteristic of toluene. These samples were then vacuum-dried and allowed to age for two weeks.

One set of salt crystals was kept as a control. Another set of crystals and the test panels were artificially aged by exposure to the unshielded light of an ultra-violet lamp for thirteen hours. (This method of aging was selected to prevent injury to the salt crystals.) After this exposure, the dammar test panel had become considerably yellowed and solubility tests on the panels indicated that some degree of aging had been achieved in both resins.

The absorption patterns of the exposed set of salt crystals were taken again after aging. The second patterns indicated that the surface characteristics had changed, causing more scattering of light, and that the chemical structure had altered slightly.[13] About 5% of the toluene, which had appeared to leave the methacrylate almost immediately, was still present in the dammar after two weeks. The controls were run. Their absorption patterns had not changed from those taken originally, except for the continued loss of toluene in the dammar control.

The aged samples were sprayed with the re-forming mixture and tests were started within a minute after each sample was sprayed. Two following results were apparent: 1) the light-scattering effect disappeared from both the dammar and methacrylate samples; and, 2) a new strong absorption band appeared immediately at 2.9 μ, where an OH group absorbs. This was, of course, caused by the OH groups in the alcohols of the re-forming solvent. The residue of toluene had disappeared from the dammar sample by the time the instrument reached the 13.7-μ band in its scanning of the spectrum, approximately fifteen minutes after the sample had been re-formed.

Repeated tests were run on the same specimens to observe the disappearance of the OH absorption band, paralleling the evaporation of the re-forming solvent. There was no evidence of the re-forming solvent in the poly(n-butyl methacrylate) specimen twenty minutes after it was sprayed with re-

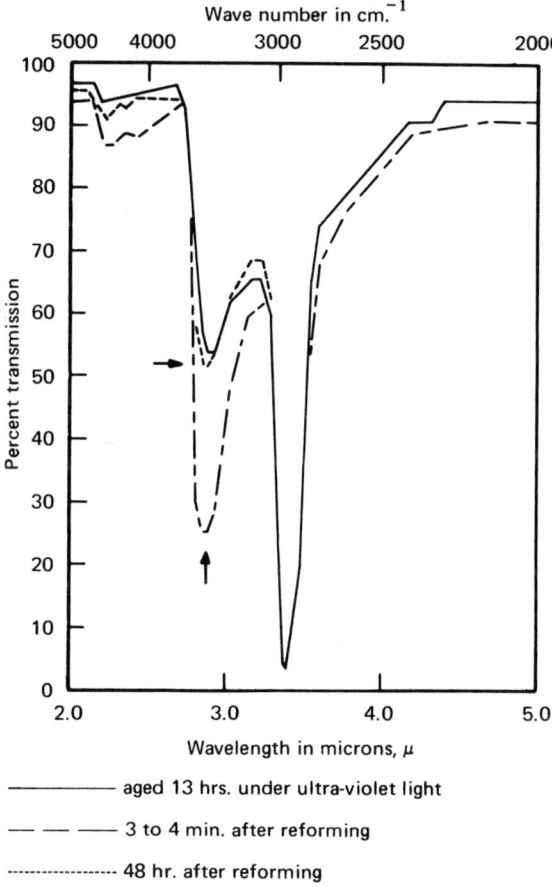

——————— aged 13 hrs. under ultra-violet light

— — — — 3 to 4 min. after reforming

·················· 48 hr. after reforming

Transmission at 2.9 μ indicates presence of alcohols.

Figure 8-2 Infra-red Spectrograph of Dammar on NaCl Prism.

forming solvent. It was estimated that about 2% of the re-forming solvent was present in the dammar after forty-eight hours. There was no evidence of any chemical change caused by re-forming (Figure 8-2).

Conclusions: Dr. Robinson concluded that infra-red spectroscopy gave no indication of major chemical changes in the resins tested.[14] In the case of dammar, the amount of re-forming solvent retained was too small (about 2%) to be measured accurately after forty-eight hours. The re-forming solvent departed much more rapidly from poly(n-butyl methacrylate). The amount retained was negligible twenty minutes after spraying. The extraordinarily rapid loss both of the original toluene used in dissolving the resin initially for coating and of the re-forming solvent aroused some interest. Mr. Doolittle had the rate of loss of toluene checked in the laboratory of Carbide and Carbon Chemical Corporation, both by a gravimetric method and by infra-red analysis. He confirmed the rapid solvent release, concluding that there is obviously a marked difference between the behavior of poly(n-butyl methacrylate) and many other film-formers in respect to solvent retention.

Radioactive Ethanol

No evidence of a chemical change was found by the tests discussed above. Can we account then for the change in solubility following re-forming by a physical change or by the influence of the very small amount of solvent retained in the film? In 1954, Dr. Arthur Solomon, Professor of Bio-Physics at the Harvard Medical School and a member of the Advisory Committee of the Department of Conservation of the Fogg Museum, came to our aid. He directed a series of tests on the retention of solvent in twenty-year-old films of four different varieties of resin through the use of radioactive ethanol. The standard re-forming solvent was used in order to duplicate the usual conditions. However, only the loss of the ethanol could be followed by this technique, since it alone was radioactive.

Certain precautions were necessary. Since the half-life of carbon 14 (C_{14}) is so long (5,770 years), the spraying of the samples was not done in the room but under a hood. Other radioactive materials were often present in nearby laboratories, thus a background count was taken before each series of tests and this count subtracted from the results; hence, the phrase "corrected counts per minute."

Specimen Material: The coatings tested had been painted out on glass in 1934. Two samples of each of the seven coatings were used. These coatings were:

1. Mastic: both prepared in the Fogg Museum laboratory
    (a) 25% in ethanol
    (b) 25% in turpentine

2. Dammar
   (a) Maximillan Toch, Toch Brothers, New York
   (b) 25% in turpentine prepared in the Fogg Museum laboratory
3. Copal
   (a) Winsor & Newton
   (b) Talens & Son
4. Vinylite A (source not recorded) prepared and painted out in 1933 in the Fogg Museum laboratory

Test Procedure: The specimens mounted on glass were cut into one-inch squares. These were attached with glue to a metal planchet which was to be placed in the Geiger counter.

The evaporation rate of the re-forming mixture was measured gravimetrically: 97-1/2% by weight of the re-forming solvent evaporated from a watch glass in fifteen minutes.

The C_{14} ethanol was diluted with normal ethanol to produce an easily measurable radiation. This diluted radioactive ethanol was used in making up the usual re-forming formula. Dr. George Hauser, who carried out these tests, had calculated the dilution to yield an initial 10,000 counts per minute from the sprayed sample, giving about 320,000 counts per minute per 1.5 ml. of re-forming solvent. (Actually, by the time the sprayed samples were in the counter and measured, the initial counts averaged less than 1000 counts per minute.)

Each sample on its planchet was placed individually under the hood in a horizontal position and sprayed with the radioactive solvent mixture.

The sample was quickly placed in the Geiger counter nearby and counting begun. The time elapse between the end of spraying and the beginning of counting was measured with a stopwatch. This operation normally required ten seconds. Time zero is the moment at which counting began.

After one minute, counting was stopped, the planchet removed from the chamber of the Geiger counter, and the count recorded. The chamber of the counter was swept with a stream of air. The sample was returned to the counter for another minute count at zero-plus-two minutes.

This operation was repeated and another count taken at zero-plus-five minutes. This count, now at a slower rate, was continued for two minutes, and the result was divided by two. Subsequent countings were continued for at least twenty-minute periods to ensure accuracy, since the number of counts per minute was smaller in each case.

Counts were taken at zero, two, and five minutes; one, four, and six days; two, eight, and fourteen weeks. The percentage of the original count remaining after each count was calculated for each sample. The averages of the specimens of each of the four kinds of resin were plotted graphically (Figures 8-3 and 8-4). (One sample each of copal, of mastic, and Vinylite A were not used in the average figures because so little solvent appeared in the first count. It is interesting to note, however, that the final corrected-counts-per-minute for these three specimens was not much less than the final average corrected-counts-per-minute of the other examples of the same resins. This indicates that, no matter how great or small was the original amount of re-forming solvent applied, the varnish tends to retain a fairly uniform amount of C_{14} ethanol after some weeks.)

Conclusions: The extreme rapidity of the initial loss of solvent came as a surprise (Figure 8-3). When the results were recorded after the first five minutes, there was doubt whether this method of testing would be successful.

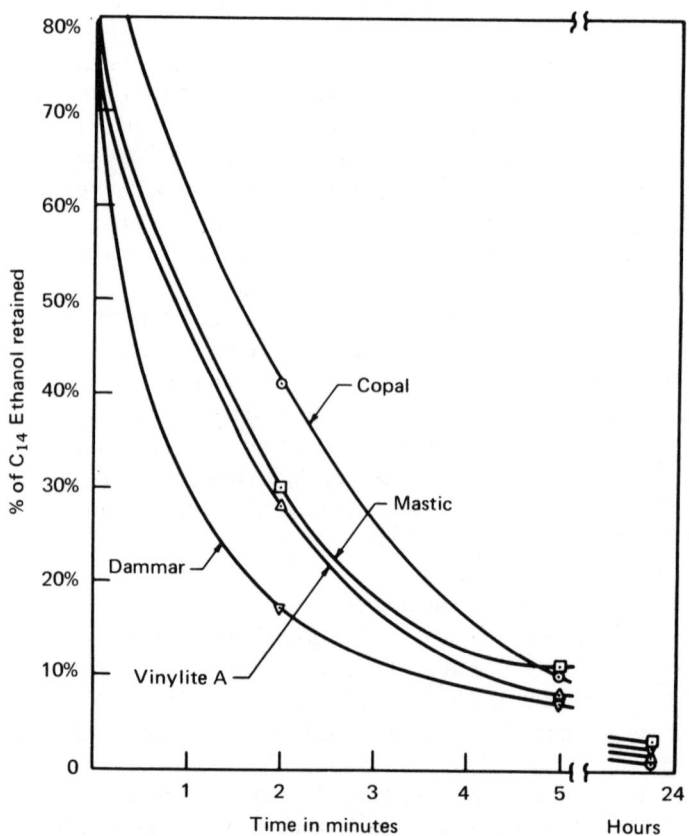

Figure 8-3 Loss of C_{14} Ethanol (in Re-forming Solvent) from Four Films Aged Twenty Years Measured by Geiger Counter.

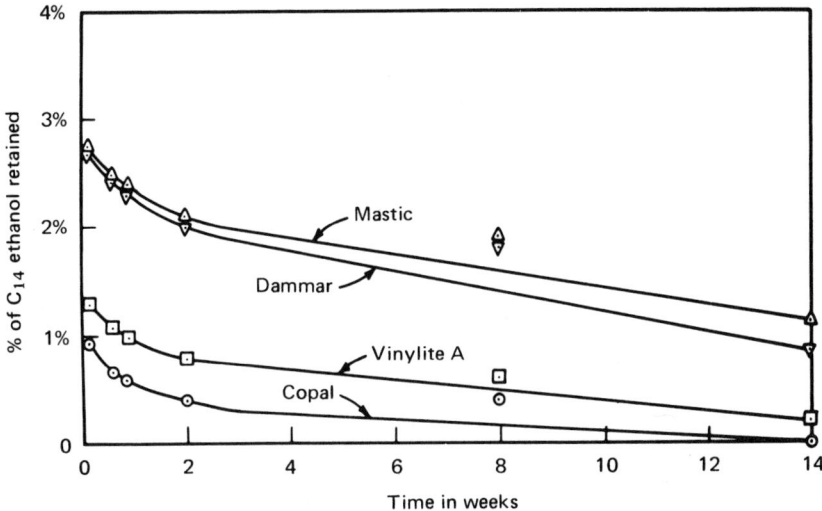

Figure 8-4 Loss of C_{14} Ethanol (in Re-forming Solvent) from Four Films Aged Twenty Years Measured by Geiger Counter.

However, it proved to be fourteen weeks before Dr. Hauser finally decided that the counts-per-minute were too small to be of interest.

At the end of fourteen weeks, there was no trace of C_{14} ethanol in any of the three copal samples. An average of 1.2% of the amount originally recorded was still present in the three samples of mastic, 0.9% in the four samples of dammar, and 0.2% in the one sample of Vinylite A (Figure 8-4).

Data which make possible a direct comparison between the C_{14} and infra-red measurements are not available. However, a relative comparison can be made if we assume that the percentage of solvent present in dammar after two minutes, as recorded by infra-red, is equal to the average figure for dammar after two minutes obtained by testing with C_{14} ethanol. In this way we find that two later measurements by infra-red agree well with the curve of the graph of C_{14} ethanol loss (Figure 8-5).

The average counts-per-minute for the three examples of copal varnish after eight weeks was 3.2 c.p.m. per square inch, less than 0.5% of the original count. This figure at eight weeks represents about a millionth of a gram of ethanol per square centimeter.

Our advisers concluded that the small amounts of solvent retained in a film after a period of twenty-four or forty-eight hours do play some part in increasing the rate of attack by solvents used in removing a re-formed coating, but that the important factor is the *swelling* of the varnish film.

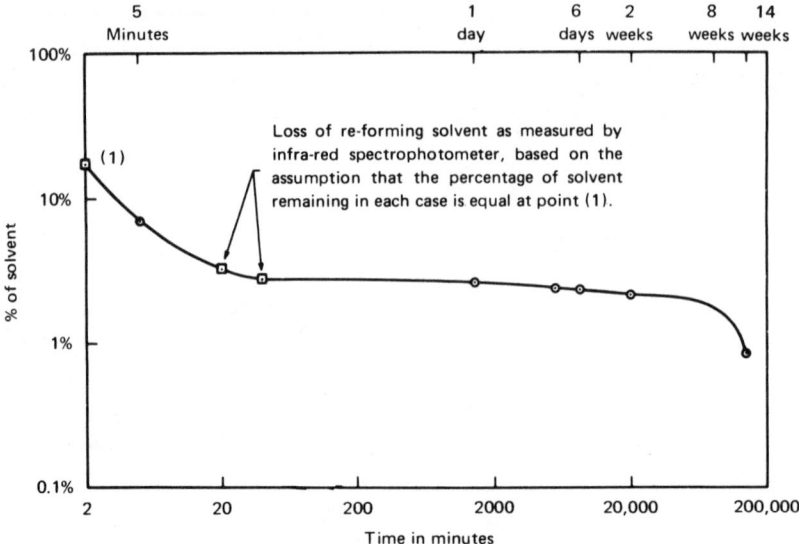

Figure 8-5 Loss of C_{14} Ethanol (in Re-forming Solvent) from Dammar Films (Average Loss from Four Specimens Aged Twenty Years).

Summary and Conclusion on the Technique of
Re-forming Varnish Coatings

We know that whatever their chemical composition, film-forming materials generally consist of molecules that are capable of uniting either by primary- or secondary-valence forces into three-dimensional aggregations of colloidal particle size. The nature, number, and location of such bonds will, to a large degree, determine the solubility, swelling, and other properties of the films so formed.

We know that when the structure of a film consists of small units or of long slender chains of molecules associated with each other only by secondary-valence forces, the film is dispersible. Thermoplastic resins are examples of such structures. Conversely, when the fundamental molecules are capable of uniting by primary-valence forces directly with other molecules to form a three-dimensional structure, the resulting film is generally insoluble. Dried oils and thermosetting resins have this structure.

The wide variation in the reaction of resins to solvents can be partially accounted for by the differences in the strength of the forces uniting the macromolecular units. X-ray diffraction studies on various resins, natural and synthetic, indicate that with increasing hardness and insolubility, the inter-molecular distances are smaller and a greater degree of regularity in the orientation of the molecules exists.[15] As resinous coatings age, they frequently become more brittle and insoluble. This is caused in part by oxidation, by the loss of certain volatile constituents in the resins, and by cross-linking between molecules. These changes often result in a greater degree of orientation of the molecules.

In re-forming, the resinous surface is sprayed briefly with a fairly "strong" solvent mixture, one in which the cohesive energy density appears to be fairly close to that of most of the resins.[16] The relatively small molecules of solvent (perhaps led by ethanol) begin to diffuse through the interstices between the aggregates of resin molecules, causing the film to begin to swell. A certain number of secondary linkages in the resin are solvated. This swelling is the first stage in the formation of the gel state of a colloid. Usually the film becomes tacky.

At this stage, the spraying is stopped. The concentration of solvent molecules begins to diminish very quickly (see Figure 8-3) by evaporation from the surface. The mass movement of solvent molecules in the film is reversed, and in due course the solvent evaporates completely.

The film, however, has been swollen and a number of secondary linkages have been broken. Dr. Paul Doty, Professor of Chemistry at Harvard University, estimates that it may take as long as six months for the original number of bonds in the aged film to be re-established.

When the process of removing a resinous coating with solvents is begun, following re-forming, the molecules of the applied solvent are better able to slip into the interstices of the opened-up film. The process of solvation is quicker and more efficient, and the varnish film is dispersed or swollen more readily than it was before re-forming. As a consequence, the paint film beneath is exposed for a shorter time to a smaller concentration of solvent molecules. The surface coating can be removed with less risk of swelling the paint film beneath.[17]

This method is not a "magic cure-all" which solves all the complex problems involved in the cleaning of paintings. We hope, however, that the re-forming technique and our greater understanding of it can become another useful tool in the hands of a trained conservator.

REFERENCES AND NOTES

1. For purposes of this report, "rate of solubility" shall be interpreted as the rate at which a solvent causes enough dispersion of the varnish coating so that it can be absorbed on a swab and removed, either as a sol or a gel of little strength. In some cases the solvent visibly affected the varnish, causing swelling and softening. If, however, the varnish could not be

removed from the panel at the termination of the test, the varnish was graded as "insoluble" in that particular solvent.

2. The data on mastic and dammar varnishes in Figure 8-1 is based on the average of all specimens. Re-forming increased the ease of removing the spirit varnishes. However, the varnishes that contained oil were noticeably less easy to remove in acetone in 1949. The considerable resistance of one specimen of each of these two resins, the specimens containing oil, resulted in a slightly greater "average" resistance being reported in Figure 8-1 for the action of acetone on either varnish and of methanol on dammar in 1949.

3. G. Thomson, "Test for Cross-Linking of Linear Polymers," *Nature* 178 (October 13, 1956): 807; "Some Picture Varnishes," *Studies in Conservation* 3, No. 2 (October 1957): 64-79.

4. Robert L. Feller, "Cross-Linking of Methacrylate Polymers by Ultraviolet Radiation," *Papers Presented at the New York Meeting, Division of Paint, Plastics and Printing Ink Chemistry, American Chemical Society* 17, No. 2 (September 1957): 465-70.

5. Dr. Robert L. Feller has been kind enough to send the author samples of an experimental methacrylate resin "27H" (polymer of isoamyl methacrylate) painted out on glass plates. One specimen had been exposed for 200 hours in the fadeometer at 60°C. This specimen was tested by our usual procedure both before and three hours after it was re-formed. Before re-forming, the coating was not completely removable in six minutes with toluene. After reforming, it came off in twenty seconds. Acetone was immediately effective both before and after re-forming. The polymer is, of course, insoluble in methyl alcohol.

6. M. von Pettenkofer, *Über Ölfarbe und Konservierung der Gemälde-Gallerien durch das Regenerations-Verfahren* (Braunschweig, 1870).

7. G. Secco-Suardo, *Il Restauratore dei Dipinti* (3rd ed.; Milano, 1918), p. 412; U. Forni, *Manuale del Pittore Restauratore*, Le Monnier (Florence, 1866), pp. 431-32.

8. A. H. Church, *The Chemistry of Paints and Painting*, 4th ed. (London: Seeley, Service and Company, 1915), p. 354.

9. M. C. Bradley, Jr., *The Treatment of Pictures* (Cambridge, Mass., 1950), Chapter 2, Section 2:06, Chapter 11.

10. G. Thomson, "New Picture Varnishes," in *Recent Advances in Conservation*, p. 176.

11. A. K. Doolittle, *The Technology of Solvents and Plasticizers* (New York: Wiley, 1954).

12. W. S. Baird, et al., "An Automatic Recording Infra-Red Spectrophotometer," *Journal of the Optical Society of America* 37, No. 10 (October 1947).

13. For further discussion of these results, see E. H. Jones, "The Effect of Aging and Re-forming on the Ease of Solubility of Certain Resins," in *Recent Advances in Conservation*: Contributions to the IIC Rome Conference 1961, ed. by G. Thomson (London: Butterworths, 1963), pp. 79-83.
14. These conclusions have been described briefly in *Better Analysis* 2 (Cambridge, Mass.: Baird Associates, Inc., 1950, now Baird-Atomic, Inc.): 9.
15. G. F. Beal, H. V. Anderson, and J. S. Long, "Studies in the Drying Oils: XVI: X-ray Study of Some Natural and Synthetic Varnish Resins," *Ind. Eng. Chem.* 24 (1932): 1068-72.
16. N. Stolow, in his paper, "The Action of Solvents on Drying-Oil Films: Part I," *J. Oil & Colour Chemists' Assoc.* (May 1957), p. 398, gives Small's equation for finding the solubility parameter of a binary mixture of solvents, providing that the two solvents mix ideally without any net volume change. By extension, the solubility parameter of the re-forming solvent mixture has been calculated to be 11.45. Its deviation from an ideal solution has not been determined.
17. In the case of an oil-resin paint coated with an oil-resin varnish, the paint usually contains a higher proportion of oil to resin than does the surface coating. The concentration of solvent molecules applied by the re-forming spray can never be as high at the level of the paint as it is at the surface of the coating. Although re-forming causes a very limited improvement in the rate of solubility of an oil-resin surface film, even a slight improvement may be of assistance to a conservator faced with one of the most demanding tests of his skill and judgement.

APPENDICES

Appendix A

EARLY STUDIES ON THE CROSS-LINKING OF METHACRYLATE POLYMERS

Robert L. Feller

While studying the depolymerization reaction of poly(isoamyl methacrylate) in 1955, Mr. Stuart Raynolds of the National Gallery of Art Research Project observed that, in the course of heating from ten to twenty hours at temperatures of from 160 to 200°C., the polymer became insoluble in hot benzene and toluene. Reports of a paper by Shultz and Bovey[2] stating that poly(t-butyl acrylate) did *not* cross-link under the conditions of their experiments appeared in the literature at about that time; also a paper by Wall was published stating that poly(methyl methacrylate) did *not* cross-link during degradation.[3] The Research Project was prompted to investigate the effects of ultra-violet radiation, as well as heat, on other methacrylate polymers. A test panel of poly(isoamyl methacrylate) currently undergoing exposure in the fadeometer was found to be highly resistant to removal with solvents.

Shortly after this first encounter with what appeared to be cross-linking under heat and ultra-violet radiation, it was necessary for the author to leave for a visit to Europe. While there, he presented samples of an experimental varnish to a few museums where he discussed with colleagues the possibility of its cross-linking. Upon the author's return to Pittsburgh, a detailed investigation was undertaken, which eventually led to the data in Table 7-4, and to a publication in 1957.[4] Mr. Thomson of the National Gallery, London, also called the author's attention to a publication by Drinberg and Yakovlev regarding cross-linking by heat.[5]

About that time the National Gallery, London, sent several types of polymers and natural-resin varnishes to an outdoor-exposure station. Upon the return of these samples to London, Mr. Thomson reported his observation of cross-linking by letter and later published a note in *Nature* on a technique which can be used to detect cross-linking of linear polymers.[6] One of the interesting results of Mr. Thomson's exposure tests was that poly(vinyl acetate) was found to be highly resistant to cross-linking, an observation later confirmed in our investigations.[7,8]

In our laboratory, simple tests were made first of the ease of removing the varnishes with solvents following their exposure to light. This was done by rubbing with cotton swabs twisted on the end of a tweezer and dipped in solvent. The 1- to 3-mil-thick films were considered to be "resistant to removal" when the time required to dissolve or otherwise wear away the film through to its support was from 75 to 105 sec. The results of exposure of various polymers in a National Accelerated Fading Unit, type XV (National Sunshine carbon arc with Corex D filter), are summarized in Table 7-4; the samples attained temperatures of about 60°C. during exposure. We have since found that slower rates of reaction and a greater degree of chain breaking will usually occur at lower temperatures (see Appendix C).

Discovering these qualities, we immediately had to ask: what do these severe conditions mean in terms of "ordinary" exposure; is it possible that such materials will change their solubility noticeably when exposed under the usual gallery conditions? From data in the Harrison Report of the Metropolitan Museum of Art, it was possible to make a correlation between the rates of fading of dyes in fadeometer tests and in actual gallery exposures.[9] The fadeometer exposure necessary to fade a class V dye had been published by K. Venkataraman.[10] On the basis of the effect of the intensity of light on the fading of a class V dye, it was estimated that 25 ± 10 hrs. in a fadeometer would be equivalent to about six years in a gallery without sun louvers, eighteen years in a gallery with sun louvers, properly operated, and eighty-two years for incandescent illumination of a level of 60 footcandles. If one assumes that the rate of cross-linking will be affected in the same ratio, then it is estimated that resistance to removal of isoamyl or isobutyl methacrylate polymers in toluene might be developed in 42 to 100 years, 126 to 300 years, or 600 years when exposed under similar gallery conditions. This estimate did not suggest an immediately alarming problem, but alerted us to give serious study to the matter. Letters were sent to colleagues in April, May, and July 1956, discussing the implications of the phenomena.

The next question to be considered was: is it possible to detect whether or not such a change has taken place within a polymer on a surface of a painting? Thomson published one method for doing this, employing a paper

TABLE A-1. TESTS OF VARIOUS POLYMERS BY THE
METHOD OF LEACHING ON PAPER

METHACRYLATE POLYMER	AGE IN YEARS	RELATIVE R_F
n-Butyl, Isoamyl, 1.5 to 3.0 mg. of Sample		1.00
Isoamyl, 8cp.* compared to 23 cp. viscosity grade		1.06
Isobutyl, 18 cp. compared to 55 cp. viscosity grade		1.08
n-Butyl (Fogg)	10	0.84 ± 0.05
Isobutyl/n-Butyl co-polymer (Fogg)	10	0.79 ± 0.01
Isoamyl (Mellon Institute)	4	0.95 ± 0.03
Isobutyl (Prof. Sheldon Keck)	5	0.93 ± 0.02
Isoamyl; resistant to removal in cyclohexane		0.50 ± 0.10

*Centipoises

chromatographic technique. At Mellon Institute, a slightly different method was developed. We hoped that paper chromatography might indicate whether or not a polymer had become of higher average-molecular-weight owing to cross-linking; data on polymers of different viscosity grade, shown in Table A-1, suggest that the rate of migration is lower for polymers of high average-molecular-weight. (The length of the migration streak in exposed samples has since been shown to be related to the percentage of insoluble matter in the sample [which influences the ease of leaching from the origin], rather than to a change in the size of the molecules which migrate up the paper.) Strips of 1 1/2" wide Whatman No. 1 filter paper were used and the eluting liquid was a mixture of equal volumes of xylene and a non-aromatic naphthenic petroleum of boiling range 170 to 190°C., density at 70°F., 0.81 g./cc. Samples were taken from the test panels on solvent-moistened swabs and applied as a spot on a pencil line marked at the bottom of the filter paper strip. The paper was then placed in a sealed chamber with this end of the strip in the eluting liquid. The liquid migrated up the paper much as it would travel up a wick, reaching a height of 30 to 40 cm. above the liquid surface after about five hours. Polymer samples of 1.5 to 3.0 mg. generally "developed" in a streak as they were swept up the paper by the eluting liquid (see Figure A-1).[11] A streak was expected both because of the delay in leaching away the sample, and because of the wide variety of molecular sizes present, each migrating at a different rate along the paper. (In the case of dammar and mastic resins, separate spots, rather than a streak, can be expected in their chromatographs, corresponding to different components in these low-molecular-weight resins.)

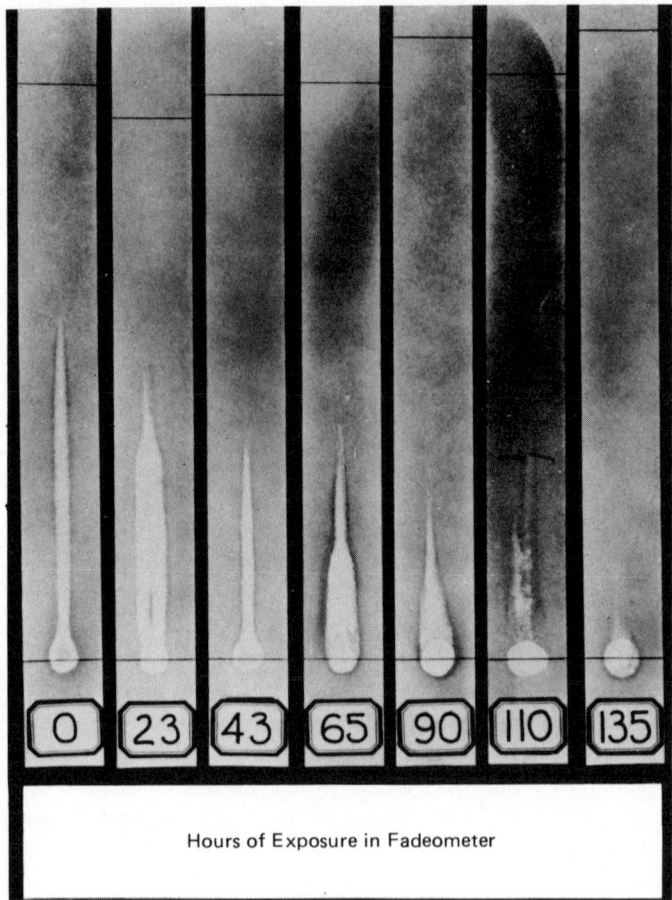

Figure A-1 Change in Length of Streaks on Chromatographic Paper with Increasing Time of Exposure of Poly(isoamyl methacrylate) to Ultra-violet Radiation.

Since this streak of polymer effectively "varnishes" the paper, the polymer was revealed by spraying the dried paper with an aqueous solution of dye. The distance from the tip of the streak of "varnish," Y, is compared to the distance traveled by the front of the eluting liquid, X, and expressed by a ratio characteristic of the material analyzed $R_F = Y/X$.

This technique differs from Thomson's in that he uses a good solvent for the polymer, sweeping all but the insoluble, cross-linked, portion away. He also adds a dye directly to the eluting solvent. In our method, we have observed that the streak becomes shorter with increased exposure to ultra-violet radiation; Figure A-2 shows a curve of R_F values relative to unexposed polymer, plotted against the time of exposure of the polymer (R_F = distance resin moved/distance solvent moved).

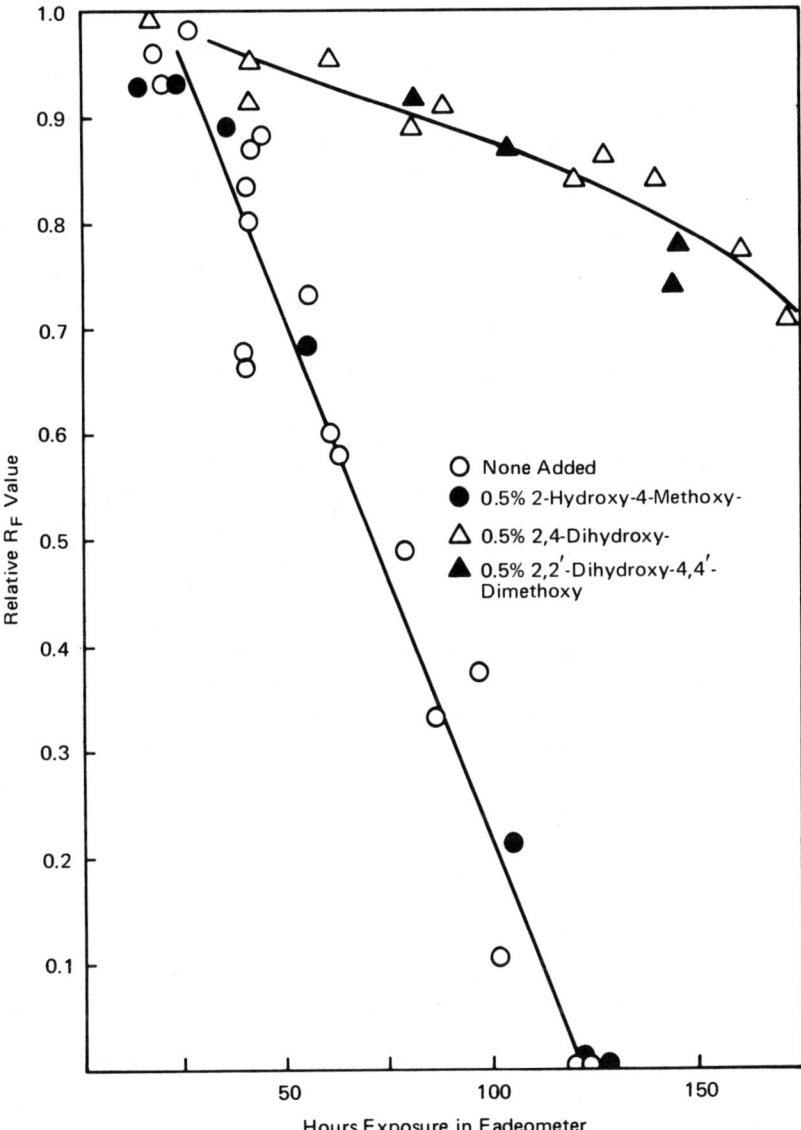

Figure A-2 *Effect of Adding Substituted Benzophenones to Poly(isoamyl methacrylate).*

Following the development of this empirical method, the Research Project sought samples of polymers which had been exposed for long periods of time indoors. At the Fogg Museum of Art at Harvard, R. J. Gettens and George L. Stout had prepared samples of various types of surface coatings and placed them on the wall of the laboratory years ago. By October 1956, many of these had been exposed for ten or fifteen years, constituting a truly valuable set of exposure panels. (For the results of solubility tests on these same panels in January 1969, see Table 7-5.) Table A-1 summarizes the results of a number of determinations on these and other polymers having considerable exposures indoors. The significant result of these studies is that the relative R_F values are not as small as they would be if appreciable degradation had occurred.[12]

If we may assume that a low relative R_F of 0.91 for five-year-old panels of isobutyl and isoamyl polymers, or 0.78 for the ten-year-old coatings, represents an actual change in the properties of these polymers, and that the decrease in the relative R_F will proceed linearly, the results imply that these polymers will reach a state in which the value of the relative R_F is 0.5 after about twenty-three to twenty-eight years of exposure similar to that which has already taken place; the latter was the figure estimated and published in 1957.[4] (At a value of 0.5 the resistance to removal with cyclohexane will become apparent when using solvents on a swab. At higher values of relative R_F, the conservator might not notice that a change of any type had taken place within the film.)

Our studies of the possibility of cross-linking up to this point in 1959 indicated that it could occur under conditions of exposure in the gallery, but that a change in the ease of removing polymers might be imperceptible for as long as twenty-eight years, unless special tests such as paper chromatography are used. In testing the behavior of poly(n-butyl methacrylate), the methacrylate most widely used in new picture varnishes, we were able to report that it did not tend to become difficult to remove with solvents as readily as the methacrylates of the isoamyl or isobutyl type. This is perhaps due to its greater ability to yield a soft swollen jelly.

Our investigation of the problem has led us to substances that do not possess this property or, at least, to ones that possess it to a lesser degree than that exhibited by isoamyl or isobutyl polymers. The high resistance to cross-linking of a co-polymer of ethyl methacrylate and methyl acrylate (Acryloid ® B-72) was reported in 1963. In addition, a number of substances related to 2,4-dihydroxybenzophenone were shown to inhibit the cross-linking tendency of methacrylate polymers. Although we have never been sufficiently satisfied with these particular compounds to recommend their use, we believe that similar ultra-violet absorbers and oxidation inhibitors can play a useful role in the future.

REFERENCES AND NOTES

1. R. L. Feller, taken from Chapter 7 of the first edition.
2. "Effects of Radiation," *Chem. and Eng. News* 33 (1955): 4051-52; A. Shultz and F. Bovey, "Electron Irradiation of Polyacrylates," *J. Polymer Sci.* 22 (1956): 485-94.
3. "How Does a Polymer Cross-Link? Effect of Oxygen May Be One Complicating Factor in Cross-linking and Degradation of Polymers by Irradiation," *Chem. and Eng. News* 33 (1955): 3390.
4. R. L. Feller, "Cross-linking of Methacrylate Polymers by Ultraviolet Radiation," *Papers Presented at the New York Meeting, Division of Paint, Plastics and Printing Ink Chemistry, American Chemical Society* 17, No. 2 (September 1957): 465-70.
5. A. Y. Drinberg and A. D. Yakovlev, "Transformation of Polybutyl Methacrylate into a Tridimensional Polymer," *Zhurnal Prikl. Khim.* 26 (1953): 532-37; *Chem. Abstr.* 47 (1953): 10318; *IIC Abstracts*, 513.
6. G. Thomson, "Test for Cross-linking of Linear Polymers," *Nature* 178 (1956): 807.
7. G. Thomson, "Some Picture Varnishes," *Studies in Conservation* 3 (1957): 64-79; see also G. Thomson, "New Picture Varnishes," in *Recent Advances in Conservation* (London: Butterworths, 1963), pp. 176-84.
8. R. L. Feller, "New Solvent-type Varnishes," in *Recent Advances in Conservation* (London: Butterworths, 1963), pp. 171-75.
9. L. S. Harrison, *Report on the Deteriorating Effects of Modern Light Sources* (New York: The Metropolitan Museum of Art, 1954); "An Investigation of the Damage Hazard in Spectral Energy," *Illuminating Eng.* 49 (1954): 253-57.
10. K. Venkataraman, "Fastness Tests and Standards," *The Chemistry of Synthetic Dyes* (New York: Academic Press, 1953), p. 297.
11. A curve showing the relationship of sample size to the observed R_F appeared in *Application of Science in Examination of Works* (Boston: Museum of Fine Arts, 1959), p. 68.
12. We cannot say positively that a value of 0.95 or even one of 0.8, represents a change of precisely 0.05 or 0.2 from the condition of the original. The reproducibility of R_F values is not precise and there also appears to be an "induction" period before any change in R_F takes place. Nevertheless, these results show that the polymers had not changed very much in five to ten years.

Appendix B

SOLUBILITY AND REMOVABILITY OF AGED POLYMERIC FILMS

Robert L. Feller

The conservator is confronted with the practical problem of removing an aged film of varnish. In discussing this problem, a clear distinction must be made between the ability to remove a polymer film with ease and safety and the inability to dissolve it.

This appendix to the original edition presents information, obtained since the seminar at Oberlin, that clearly illustrates the various states or stages that a coating passes through when the polymer predominantly cross-links upon aging. Four stages may be visualized: a period of induction; a period of cross-linking until practically all the film is insoluble; a period of increased degree of cross-linking until the film is no longer swollen by solvents; and, finally, a period of non-swelling.

Appendix C will describe subsequent studies on the behavior of polymers that undergo a significant amount of chain breaking as well as cross-linking.

INDUCTION PERIOD

Figure B-1 illustrates the increase in the percentage of insoluble material in the experimental varnish 27H,[1] upon exposure, at a temperature of about 60°C., to ultra-violet radiation in a fadeometer. Elvacite ® 2044 poly(n-butyl methacrylate) behaves in a similar manner. A period of about twenty-six hours is seen in Figure B-1, in which no loss of solubility occurs. This may be considered to be a period of *induction* during which the molecules of the colloidally dispersed *sol* are growing to sufficient size that they

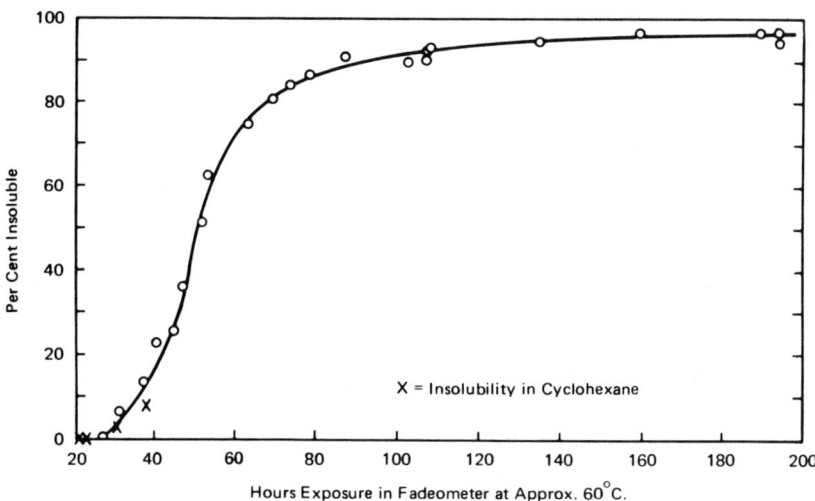

Figure B-1 Percentage of 27H Varnish Insoluble in Toluene at Room Temperature, After Exposure on Aluminum Foil in the Fadeometer.

will eventually result in a *gel* (insoluble).[2] The tendency to grow can be *inhibited* by the presence of various substances in the film, either naturally present, or intentionally added.

PERIOD OF DECREASING SOLUBILITY

A second period may be visualized in Figure B-1, ranging from about twenty-six to eighty hours, during which the polymer develops insolubility owing to the formation of cross-links. The method of determining the insolubility reported in Figure B-1 was similar to the determination of gel-content described by Shultz and Bovey; in one way it defines insolubility, that is: the inability of the swollen gel to pass through a 120-mesh/in. screen.[3,4]

Only a small number of cross-linked chemical bonds is necessary before a linear polymer, of the degree of polymerization of 100 or greater, becomes insoluble.[5] Boyer, for example, has shown that the presence of only about 0.01% by weight of divinylbenzene (cross-linking agent) in a polystyrene leads to the formation of an insoluble gel.[6] Insolubility occurs when the molecules become very large; the condition of insolubility, therefore, is little influenced by the solubility parameter of the solvent involved. Several points marked by "X" in Figure B-1 show that the insolubility of the film occurs at about the same time both in cyclohexane and in toluene.

In this second stage, a portion of the film is still soluble; the film may be

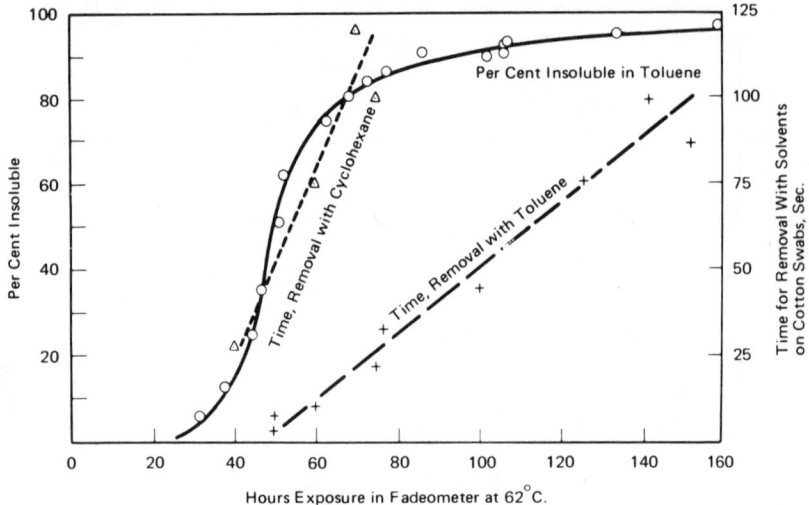

Figure B-2   Behavior of 27H Varnish, Coated on Aluminum Foil, Following Exposure in Fadeometer.

removed with considerable ease as a mixture of sol and gel. The data in Figure B-2 indicate that resistance to removal in cyclohexane, a solvent in which polymer 27H was initially soluble, increases with increasing insolubility. It has also been noted that, in the Research Project's technique of leaching-on-paper to measure cross-linking, the decrease in relative R_F value followed closely the increase in insoluble matter in 27H varnish, becoming zero when the insoluble portion amounted to at least 90% of the film.[7]

To determine whether a linear polymer has entered this stage in the process of cross-linking, one need only detect the presence of polymeric material in the film that will not dissolve in the best of solvents. Thomson has described a simple method of leaching-on-paper that does just this.[8] When poly(n-butyl methacrylate) and poly(isoamyl methacrylate) were tested by his method, we found that material remained at the origin spot even when 75% of the film was able to pass through the 120-mesh screen; in other words, the method developed by Thomson detects the presence of insoluble matter within the polymer film even though a considerable fraction still dissolves.

Thomson's test is admirably simple and sensitive for the detection of the very first stages of cross-linking, and may be called a qualitative test. To obtain a quantitative measure of the extent of insolubility, the fraction of soluble or insoluble matter may be determined directly. Thomson accomplished this by weighing a small chip of varnish; we make a similar analysis by taking up the varnish on weighed cotton swabs, drying them, and then extracting the varnish. The Research Project's empirical method of leaching-on-paper (Appendix A) was an attempt to follow the course of insolubilization indirectly; it was useful at the time but is no longer employed.

As the following appendix will point out, the rate at which the insoluble matter builds up in time follows a mathematical law, usually becoming about 90% insoluble after five times the "gel dose" (the exposure necessary before the first insoluble matter is formed). If appreciable chain breaking occurs simultaneously, the rate of buildup of insoluble matter will not occur as rapidly.

PERIOD OF DECREASING SWELLING

The degree of swelling is controlled in part by the chemical nature of the solvent and polymer, but it is particularly interesting that, for a given system, the swelling is controlled by the degree of cross-linking. As cross-links multiply and the average number of monomeric units between each cross-link becomes smaller, the ability to swell lessens. Based on research of the late 1940s, Boyer has reported that the swelling of cross-linked polystyrene in toluene and methyl ethyl ketone becomes markedly reduced as the weight per cent of cross-linking agent, divinylbenzene, is increased from 0.02 to 0.14%.[6] The swelling of polybutadiene in carbon tetrachloride is known to decrease as the time of heating at 255°C. is increased.[9] In more precise studies, Eirich has demonstrated that one glycol dimethylacrylate cross-link in every one hundred units of a methyl methacrylate chain results in a polymer that swells to about three times its original volume.[10] Based on the fundamental equations of Flory and Rehner, it can be shown that, when the average number of monomer units between cross-links in any polymer falls below 10, a polymer usually will not swell more than twice its original volume even in the most effective solvent.[11]

We have shown in Figure B-2 that the time required to remove 27H varnish with a cotton swab dipped in cyclohexane is closely related to the amount of insoluble matter present. On the other hand, it is apparent that it is *not* the percentage of insoluble matter that influences the ease of removal with toluene. The resistance to removal with toluene and acetone at this later stage apparently is influenced by the degree of swelling, which is dependent upon the degree of cross-linking. Polymers of isoamyl and n-butyl methacrylate, when exposed on aluminum foil in a fadeometer at about 60°C., were found to become 70 to 80% insoluble in about sixty hours; Table B-1 shows that the resistance of these polymers to removal in toluene and acetone does not develop until after many more hours of exposure. These observations clearly indicate that, even though a film may be almost completely insoluble, it may still be removed by solvents as a swollen gel with reasonable ease until cross-linking has progressed to a much-advanced stage, at which point the cross-link density has become very high.[11] In 1963, the author published further data to demonstrate this point.[12]

TABLE B-1. HOURS OF EXPOSURE ON ALUMINUM FOIL IN FADEOMETER AT 60°C. BEFORE DEVELOPING RESISTANCE TO REMOVAL IN VARIOUS SOLVENTS*

METHACRYLATE POLYMER	VISCOSITY GRADE OF POLYMER	CYCLOHEXANE	TOLUENE	ACETONE
Isoamyl....	23	70	131	358
n-Butyl....	48		295	580

*Resistance-to-removal is defined in Table 7-4, footnote (a). The times in Table 7-4 are to be divided by 1.9 to compare them to the exposures on aluminum foil reported here and in Figures B-1 and B-2 in this appendix.

PERIOD OF NON-SWELLING BY SOLVENTS

A fourth stage can be envisioned in which a polymer has become so highly cross-linked that it is not swollen to any practical degree by solvents. We know little about this highly advanced stage in the aging of films. However, the removal of such a film would be expected to be very difficult, unless by chance or design, it may be sufficiently brittle that it can be removed by pulverization. At this stage, deterioration reactions are probably taking place in which the polymer undergoes degradation, the chains, or parts of the chain, breaking and becoming once again partially dispersible. The evolution of volatile degradation products, for example, has been observed throughout periods II and III in the deterioration of poly(n-butyl methacrylate) and poly(isoamyl methacrylate) [Table 7-6] and may be expected to continue in period IV.

SUMMARY AND CONCLUSION

Four periods can be envisioned in the aging process of a linear polymer that tends to cross-link excessively; Figure B-3 summarizes the situation. The implications in the problem of removing a varnish are as follows: after a period of induction, 1, the film may begin to become insoluble and resistant to removal in solvents in which it was originally completely soluble. Following this second period, we may consider a third period, in which the film is almost totally insoluble in solvents, but may still be swollen (although solvents of higher solubility parameter than those required by the original polymer may be needed if the material oxidizes). The ability to swell decreases as the degree of cross-linking increases; nevertheless, until this be-

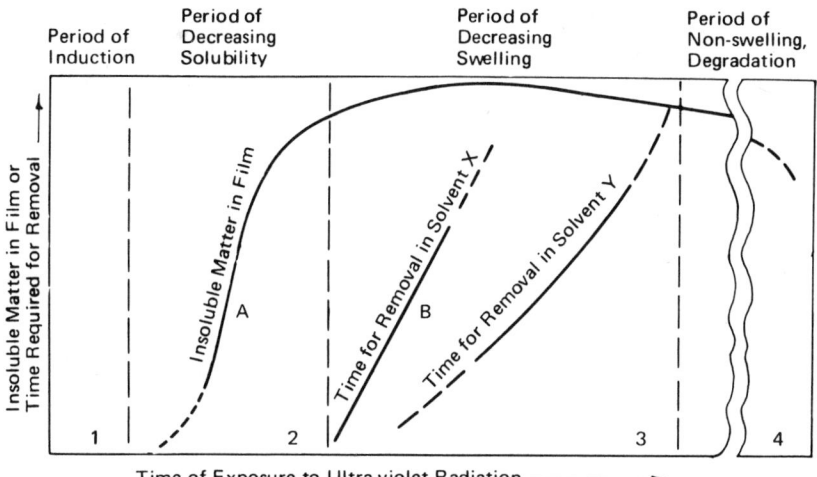

Figure B-3 Schematic Representation of the Changes in the Ease of Removal in Solvents of a Film of a Linear Polymer That Undergoes Cross-linking.

comes extensive, it is possible to remove the film as a swollen gel. Beyond this, one may visualize a state, 4, in which the film is unaffected by solvents and is extremely difficult to remove.

As a chemist investigating the properties of protective coatings, the author would prefer that no picture varnish proceed to a state beyond point A in Figure B-3, perhaps in the region of 25 to 50% insoluble matter.[13] Tests are available to detect and measure the extent to which a polymeric film has progressed in stage 2. Chromatographic tests of poly(n-butyl methacrylate) samples on a well-lighted wall of the Fogg Art Museum (Appendix A), for example, showed that after ten years the polymers had only just entered stage 2. Although an accurate prediction could not be made at such an early stage in 1957, it was estimated on the basis of this test that point A probably would not be reached in these samples for about twenty-eight years. Several other tests also indicate that the minimum time should be of this order in galleries well-illuminated by daylight. Because this time was sufficiently great, there seemed little cause for alarm in 1957-1959. However, it was short enough that we felt the subject should be thoroughly investigated. Solubility tests run in January 1969 (Table 7-5), show that, after twenty-two years, samples on the wall of the laboratory of the Fogg Art Museum have reached about point A.[14]

Even after the film of polymer has passed through state 2 and has become almost completely insoluble, it may be removed with considerable ease as a gel during an additional period of exposure. Removal with solvents, however, becomes more difficult as the degree of cross-linking progresses. The author would prefer that a varnish would not pass beyond point A in Figure B-3, but would certainly consider it to be most serious if a picture varnish made with a polymer should progress to a state beyond point B. This point may be defined arbitrarily as a state requiring more than 105 seconds of rubbing with a cotton swab dipped in toluene in order to remove the insoluble film.[15] As Table B-1 shows, this condition is not expected to arise for a period of two or four times longer than necessary to reach point A. Even then, when they had developed resistance to removal in toluene, the films in Table B-1 could still be removed as a gel in another solvent.

The investigations just described enabled us in 1959 to characterize in detail the changes in solubility and in ability to swell that could be expected to take place in films of thermoplastic polymers that tended to cross-link extensively upon aging. This information, gained on the basis of "accelerated aging tests," enabled conservators to avoid inappropriate applications of butyl and isoamyl methacrylate polymers long before their use was able to cause difficulties. Moreover, a thorough understanding of these processes is fundamental to the proper utilization of thermoplastic resins in conservation.

Since these first studies were published, we have also been able to detect significant changes in weight and solubility that may take place during aging. In addition, we have investigated the situation in which a significant amount of chain breaking, as well as cross-linking, may occur. The results of these subsequent studies are presented in Appendix C.

REFERENCES AND NOTES

1. Poly(isoamyl methacrylate) of 23 cp. viscosity grade dissolved in non-aromatic, naphthenic petroleum, b.p. 170-210°C., density 0.81 at 25°C.
2. A curve showing the region of induction and subsequent increase in gel content in a typical three-dimensional polymer can be found in P. J. Flory, *Principles of Polymer Chemistry* (Ithaca, New York: Cornell University Press, 1953), p. 382.
3. A. R. Shultz and F. A. Bovey, "Electron Irradiation of Polyacrylates," *J. Polymer Sci.* 22 (1956): 485-94.
4. Samples of the exposed films, peeled free or left on aluminum foil, were weighed, placed in 120-mesh/in. stainless steel cages, and shaken in toluene for forty-eight hours at room temperature. (A higher temperature usually results in the extraction of a slightly greater amount of material.) Drying of the extracted resin was done at 70°C. for twenty-four hours.

The initial weight of the resin usually was about 30 to 50 mg., although the method has been used successfully on only a few milligrams removed on previously weighed cotton swabs.

5. Noted by G. Thomson, "Some Picture Varnishes," *Studies in Conservation* 3 (1957): 64-79, footnotes 1, 2.
6. R. F. Boyer and R. S. Spencer, "Some Thermodynamic Properties of Slightly Cross-linked Styrene Divinyl Benzene Gels, II. The Swelling Characteristics of Slightly Cross-linked Gels," in H. A. Robinson (ed.), *High Polymer Physics* (Brooklyn, New York: Remsen Press, 1948), p. 465.
7. Between forty and seventy hours, the leaching-on-paper method was found to give erratic results. With the marked rise in insoluble material, it is difficult to control the amount and nature of the sample at this time; and, hence, erratic results are to be expected.
8. G. Thomson, "Test for Cross-linking of Linear Polymers," *Nature* 178 (1956): 807.
9. J. A. Coffman, "Highly Cross-linked Polybutadiene, Preparation and Mechanical Behaviour," *Ind. Eng. Chem.* 44 (1952): 1421-28.
10. F. R. Eirich and R. Lauria, "Model Experiments on the Re-wetting of Chalked Paints," *Official Digest, Federation of Societies for Paint Technology* 32 (1960): 183-95.
11. A. K. Doolittle, "Theory of Solvent Action of Swelling," *The Technology of Solvents and Plasticizers* (New York: Wiley, 1954), p. 830; P. J. Flory, "Phase Equilibria in Polymer Systems; Theory of Swelling," *Principles of Polymer Chemistry* (Ithaca, New York: Cornell University Press, 1953), p. 580.
12. R. L. Feller, "New Solvent-type Varnishes," in *Recent Advances in Conservation* (London: Butterworths, 1963), pp. 171-75.
13. The traditional mixture of dammar plus 5% stand oil by volume would yield a film containing about 14% stand oil by weight. Stand oil, of course, cross-links and there is considerable experience with the removal of such films.
14. The age of the samples at the Fogg Art Museum is now about twenty-two years. A proprietary varnish based on poly(n-butyl methacrylate) has been available in America for about thirty years. These are close to the maximum age one might expect to find in a well-documented varnish based on a methacrylate polymer. In tests made on twenty-two-year-old films of n-butyl methacrylate films from the Fogg test samples, it was found that all were easily removed in mixtures of 50/50 methyl cyclohexane/xylene and were no more than 57% insoluble. Exposure and tests on these samples will continue.

15. Aged dammar varnish usually requires a solvent of higher solubility parameter than toluene for its removal, but it usually does not require the rubbing, the mechanical action, that an increasingly tough gel of polymer may be expected to need for its removal.

Appendix C

STUDIES OF THE EFFECT OF LIGHT ON PROTECTIVE COATINGS USING ALUMINUM FOIL AS A SUPPORT: DETERMINATION OF RATIO OF CHAIN BREAKING TO CROSS-LINKING[1]

Robert L. Feller

The deterioration of protective coatings is usually hastened by high-intensity illumination and a high proportion of ultra-violet radiation, although the ultimate aim is to predict behavior at much slower rates of deterioration under normal conditions. When our laboratory first began the study of the aging of methacrylate polymers in a carbon-arc fadeometer, we coated the films on sheets of aluminum foil at the suggestion of colleagues on the Bakelite Fellowship at Mellon Institute. The rate of deterioration of the coatings is effectively increased with the use of foil because the radiation must pass through the film twice, once as it enters the film and again as it is reflected back by the aluminum. Earlier, we reported that the increase in the rate of cross-linking was approximately a factor of 1.9, nearly twice as fast on foil as when the coatings were placed on ordinary window glass.[2]

The use of aluminum foil has an additional advantage in that samples can be readily weighed. When the coatings are exposed to intense radiation, they also generally lose weight and this is easily followed by weighing.[3] Since the foil is stable, the volatile products of deterioration may be collected and analyzed, uncontaminated by extraneous organic material. During exposure in the fadeometer, samples may also be obtained periodically by cutting out pieces approximately 1/4" wide and 2" long, or larger, with a razor blade and extracting the soluble portion of the films.

More than ten years have passed since our first publication concerning the occurrence of cross-linking in methacrylate polymers. As we re-emphasized at the IIC conference in Rome in 1961, the tendency for thermoplastic polymers to develop insoluble matter upon aging is rather the more general rule than had hitherto been expected.

Insolubility of a thermoplastic coating in a given solvent can arise not only through cross-linking, but also through the polymer becoming partially oxidized and changing its solubility parameter. In the case of Acryloid ® B-82 and Rhoplex ® AC-33, for example, increased insolubility in toluene with age cannot be interpreted as being due to cross-linking, but apparently, to a change in the solubility parameter. By changing from toluene, the usual solvent, to one of higher solubility parameter, acetone, the aged films are effectively dissolved.[4] This important aspect of the aging of polymers is another phenomenon that was demonstrated by the simple technique of weighing and extracting films coated on aluminum foil.

At the Rome meeting in 1961, it was pointed out that light could have a number of effects on polymers. In addition to a weight loss and the change in solubility characteristics, a major effect involves the breaking of polymer chains as well as joining up (cross-linking). Indeed, if a significant amount of chain breaking as well as cross-linking takes place, the detection of cross-linked matter within an aged thermoplastic coating would not necessarily mean that the film would eventually become completely insoluble. Although the subject of a simultaneous chain breaking and cross-linking was discussed in detail at the Rome meeting, representing our latest laboratory results, the original paper was not changed to include this new interpretation of the results. We take this opportunity now to show how this further significant factor in film deterioration, chain breaking, can be demonstrated by simple means.

The rise in insoluble matter in films of three methacrylate polymers is shown in Figure C-1. The method used to determine this was the same as that employed when such data were first reported by us (Figure A-1). At various times during the exposure in the fadeometer, strips of the coating on aluminum foil were cut with a razor, weighed, placed in fine-mesh wire baskets, and extracted in solvent. Following extraction, the films were dried and

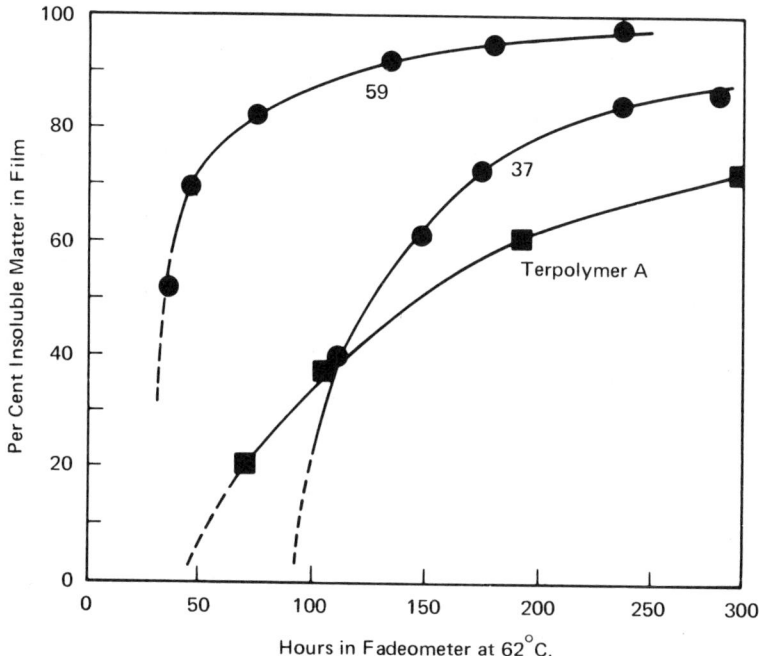

Figure C-1 Increase of Insoluble Matter in Several Thermoplastic Polymers upon Exposure in Fadeometer.

weighed again to determine the remaining insoluble material. The weight of the foil was determined finally by mechanically removing the solvent-swollen insoluble film from the foil.

In Figure C-1, polymer 37 is poly(isobutyl methacrylate); polymer 59 is a co-polymer of p-methylcyclohexyl and isoamyl methacrylate previously reported by this laboratory to have a marked tendency to cross-link (Table 7-4), and terpolymer A is a commercial acrylic resin.

In these studies, it was soon noticed that some of the data, such as the curve for the terpolymer A in Figure C-1, did not show a rapid rise to practically complete insolubility. An explanation and interpretation of this slower rate of change can be found in analysis of the problem by Charlesby.[5] Charlesby has developed equations to predict the rise in insoluble matter in a polymer that is undergoing simultaneous chain scission and cross-linking. He expresses the change in the soluble fraction of such films as a function of the time or "radiation dose" necessary to cause the initial appearance of insoluble gel, R^*. A family of curves has been calculated to show the decrease in the soluble fraction as a function of the relative time since the initial formation of gel, R/R^* (Charlesby, p. 211). The curves change in slope as the ratio of chain scission to chain-linking varies, resulting in curves similar to those in Figure C-2 (the three sets of data shown were among those presented at the meeting in Rome in 1961).

In Figure C-1, it can be seen that polymer 59 began to form an insoluble fraction at the earliest time; the formation of insoluble matter also rises faster than in the other polymers. Polymer 37 required a longer time of exposure

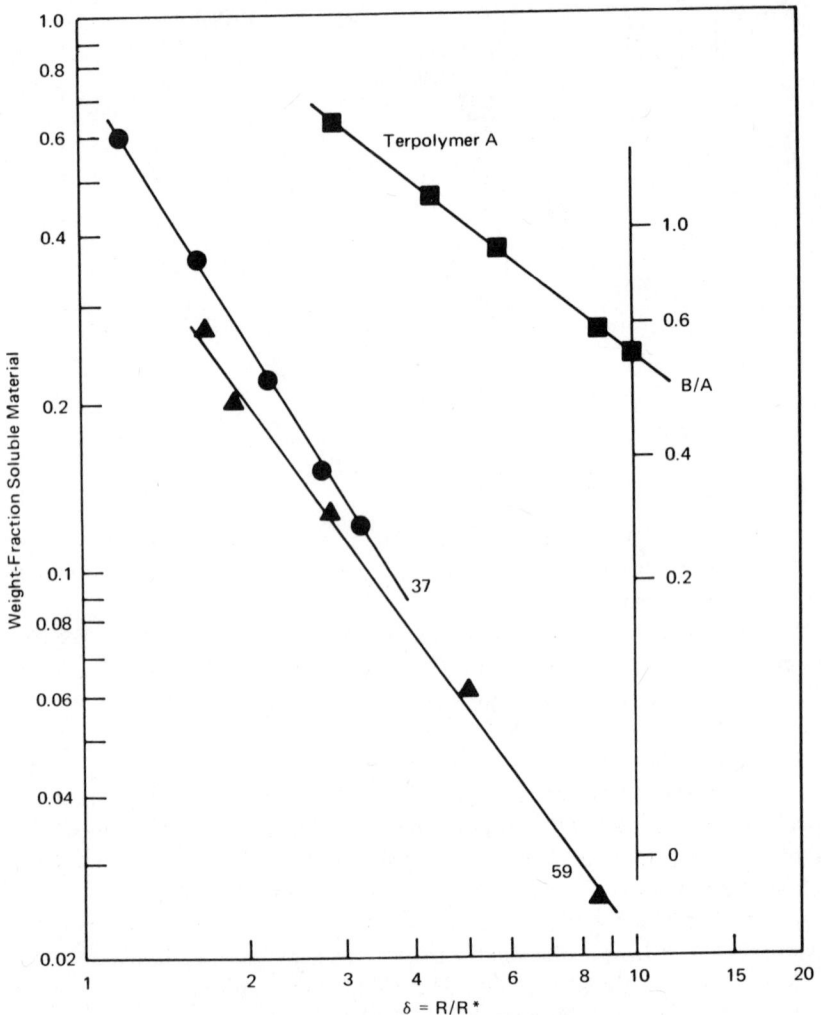

Figure C-2 Decrease in Soluble Fraction of Films in Relation to the Relative Time Since Initial Formation of Gel.

before the initial formation of insoluble matter. According to Charlesby's analysis, the increase in insoluble matter for polymer 37 in Figure C-1 should occur at a slower rate than that for polymer 59, as indeed it does. When the data from Figure C-1 are plotted in Figure C-2 with respect to R/R^*, according to Charlesby's recommendations, it may be seen that the data for methacrylate polymers 59 and 37 are practically identical.

In Figure C-2 the decrease in the soluble fraction in terpolymer A does not proceed at the same rate as in the other two polymers. As already mentioned, Charlesby uses a family of curves to show that the soluble fraction may not fall as rapidly as experienced in polymers 59 and 37, but will fall at different rates according to the amount of chain breaking B, and cross-linking A, taking place simultaneously. The family of curves is not drawn here but only the position that they would reach after a period of ten times the initial gel dose, $R/R^* = 10$. The amount of soluble material that would remain in the film at this point can thus be taken as an indication of the ratio of the number of chain scissions that occur for every chain that is cross-linked, B/A. Terpolymer A thus is seen to exhibit a ratio of the number of polymer chains broken to the number of cross-links formed of about 0.55. On a basis of this analysis, it can also be seen that polymers 59 and 37 cross-link almost exclusively with no significant amount of chain breaking taking place. Poly(n-butyl methacrylate) behaves in the same way.

The ultimate purpose of these investigations is to predict the behavior at room temperatures. In this regard, it should be pointed out that the ratio of chain breaking to chain-linking can vary with temperature and, in fact, in the case of polymers 37 and 59, the ratio increased when they were tested at a lower temperature (32°C.) [a greater degree of breaking took place].

Figure C-3 shows a more general and customary way of plotting these data. Here a plot of $S + \sqrt{S}$ is made against $1/\delta$, where S is the soluble fraction that remains in the films and $1/\delta$ is the ratio of the time to cause initial gelation to the time of exposure, R^*/R, the inverse of the unit used in the x-axis in Figure C-2. This approach is used in another analysis published by Charlesby (op. cit., p. 173) and has the advantage that the slope of the curves should generally be 2. It is difficult to determine precisely the time for initial formation of gel, based on the extrapolation of the curves in Figure C-1 to 0% insoluble matter. In the data used for the three polymers, errors were apparently made in the selection of the initial gel time because the plots of the data in Figure C-3 do not have a slope of 2 (indicated by the dotted line). Correcting these data to bring the slope to a value of 2 may be helpful in choosing the most reasonable value for the initial gel time. In this plot, the intercept at $1/\delta = 0$, gives a ratio of B/A of about 0.5 for terpolymer A, compared to 0.55 determined in Figure C-2.

Consideration of the ratio of B/A in investigations of the effects of ultra-

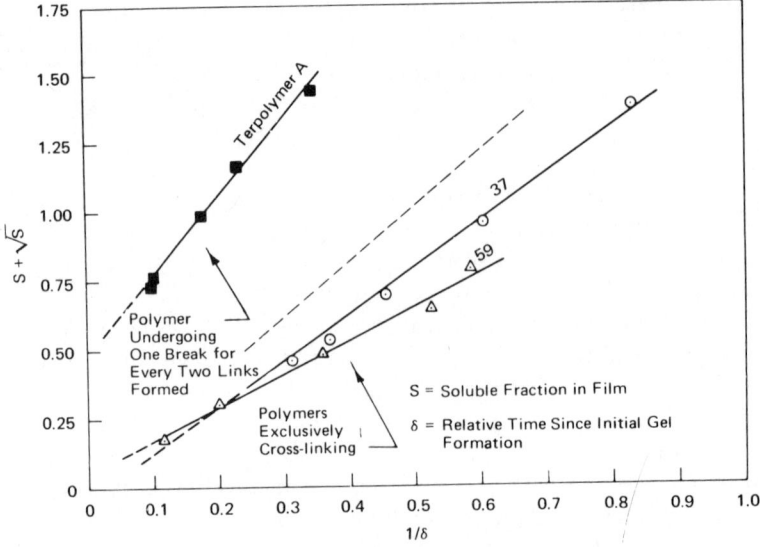

Figure C-3 Change in Soluble Matter in Polymers with Time of Exposure to Ultra-violet Radiation.

violet light on films has appeared in the literature in recent years.[6] One of the most elaborate analyses of the problem has been made by L. D. Maxim and C. H. Kuist, in which the deterioration of several methacrylate and acrylate polymers has been reported. They have extended their detailed analysis to reveal that the bonds in co-polymers that are the most readily broken can be detected.[7,8]

Certainly, the analysis of the loss of weight that takes place when a film ages, and the relative importance of chain breaking and chain-linking processes, will soon become a widespread and standard practice in the study of the deterioration of thermoplastic films.

SUMMARY

Four different processes have been shown to be taking place in the aging of methacrylate polymers: (1) weight loss; (2) change in "solubility parameter" of solvent required to dissolve the film; (3) breaking of polymer chains; and (4) cross-linking. Thermoplastic coatings exposed to ultra-violet light tend to lose weight and, later, to require "stronger" solvents for their removal. These changes are thought to occur because the polymer molecules become oxidized to some degree. Apparently because a similar oxidation process takes place in natural resins, they too usually require "stronger" solvents to dissolve them as they become older.

Molecules in a thermoplastic resin can also develop insoluble matter in a film, regardless of the solvent used, if the molecules become cross-linked in time. A process of breaking polymer molecules as well as linking can also occur; "chain breaking" of molecules is one of the causes of embrittlement of textile and paper fibers. One process or the other may predominate in certain polymers, but often both occur at the same time. If appreciable chain breaking occurs, the buildup of insoluble matter in a film will take place at a slower rate than would be the case if the molecule tended to cross-link exclusively. Laws that predict the rate at which the buildup of insoluble matter occurs in a film are precisely known, and from these the relative rates of the breaking and linking can easily be determined.

REFERENCES AND NOTES

1. R. L. Feller and C. W. Bailie, "Studies of the Effect of Light on Protective Coatings Using Aluminum Foil as a Support: Determination of Ratio of Chain Breaking to Cross-linking," reprinted with slight changes from the *Bulletin of the American Group-IIC* 6, No. 1 (April, 1966).
2. R. L. Feller, "Cross-linking of Methacrylate Polymers by Ultraviolet Radiation," *Papers Presented at the New York Meeting, American Chemical Society, Division of Paint, Plastics and Printing Ink Chemistry* 17, No. 2 (September 1957): 465-70.
3. C. J. Berg, W. R. Jarosz, and G. F. Salathe, "Performance of Polymers in Pigmented Systems," *J. Paint Technology* 39 (1967): 436-53.
4. R. L. Feller, "New Solvent-type Varnishes," in *Recent Advances in Conservation* (London: Butterworths, 1963), pp. 171-75.
5. A. Charlesby, *Atomic Radiation and Polymers* (London: Pergamon, 1960).
6. G. B. Stephanson and W. S. Wilcox, "Ultraviolet Irradiation, IV. Further Studies of Environmental Effects on Films and Fibers," *J. Polymer Sci.* 1A (1963): 2741-52.
7. L. D. Maxim and C. H. Kuist, "The Light Stability of Vinyl Polymers and the Effect of Pigmentation," *Official Digest, Federation of Societies for Paint Technology* 26 (1964): 723-44.
8. L. D. Maxim and C. H. Kuist, "The Ultraviolet Degradation of Scissioning Copolymers," *Polymer* 6 (1965): 523-29.

Appendix D

POLYMER EMULSIONS[1]

Robert L. Feller

Milky-white emulsions of polymers have become familiar materials in the home, artist's studio, and restoration workshop. They are usually purchased already made, not prepared by the conservator himself. For this reason, it is profitable to review some of their essential properties.[2]

"Polymer emulsions" consist of minute particles of polymer dispersed in water. The typical volatile organic solvents are not present; the milky fluid is primarily water and tiny globules of polymer dispersed in it. Essential minor constituents are present, however, particularly the dispersing agent used to keep the particles of polymer well-dispersed. There may also be plasticizer present in some preparations, and, occasionally, a thickening agent.[3] Polymer emulsions dry upon loss of the water through evaporation and form a satisfactorily strong film, owing to the fact that the particles of polymer touch one another and eventually flow together to form a continuous film of the polymer.

The commercial emulsions are not made by "whipping" together a mixture of polymer solution in water. Instead, they are prepared by "emulsion polymerization." In this process, the mixture of monomer and dispersing agent is dispersed into fine droplets in the water. When the dispersion is considered satisfactory, a polymerization catalyst is added, causing the droplets of monomer to polymerize. The final result is the formation of particles of polymer dispersed in water.

Thermosetting polymers are also formulated as emulsions, but the present discussion will be limited to thermoplastic films since these are related in their ultimate physical properties to the solvent-type coatings already discussed.

One of the key ingredients to consider in examining emulsions is the dispersing agent that has been used to stabilize the finely divided emulsion. Some dispersing agents cause the final emulsion to be slightly alkaline or acid.[4] If pigments are to be used with such emulsions, they must be acid- or alkali-resistant. Other dispersing agents produce practically neutral emulsions.[5,6] The conservation laboratory must check emulsions with indicator papers or pH meters before using them in applications where acidity or alkalinity may be an important factor.

One of the chief advantages of polymer emulsions is that they contain high solids, usually around 50% polymer, and yet they are not exceptionally viscous. When polymers are dissolved in a liquid, the viscosity is a function both of the concentration of the polymer present and the size of the polymer molecules (molecular weight). When a polymer is dispersed as an emulsion, however, the viscosity is determined primarily by the viscosity of the tiny polymer particles flowing past one another through the water. Hence, the viscosity is affected by the concentration of dispersed polymer droplets but is insensitive to the size of the polymer molecules inside the droplets. For this reason, it is possible to use polymers of very high-molecular weight in emulsions (high intrinsic viscosity), with the result that the final film can possess high tensile strength. When large polymer molecules are dissolved in an organic solvent, the viscosity of the solution is so very high that there is considerable difficulty in handling, and the solution usually has to be used at a relatively low polymer concentration. The reason for the great popularity of polymer emulsions is that they have the two distinct advantages described, i.e., (a) a high concentration of resin can be used; and (b) a relatively low viscosity is maintained in spite of the use of high-molecular-weight polymers.

The process of drying of a polymer emulsion is illustrated in Figure D-1. At first particles of the resin are rather widely dispersed in water. The viscosity of these particles flowing in water and past one another primarily controls the viscosity of the emulsion, rather than the entanglement of large molecules of the resin itself, as is the case in a solution of the polymer. The scattering of light from the globules of resin causes the emulsion to appear milky.

In the first step of drying, the water quickly evaporates, and the individual particles of resin are able to touch one another as shown in stage 2. Data in Table D-1 show that, in a few hours, 97% of the water has evaporated and, after one week, nearly all of the water is gone. The properties of the film represented by stage 2 are not the ultimate properties of the polymer itself. Very often, the coating at this stage can still be softened and dispersed in water.

The process by which the coating changes in Figure D-1 from stage 2 to stage 3, in which the droplets have flowed together and formed a practically

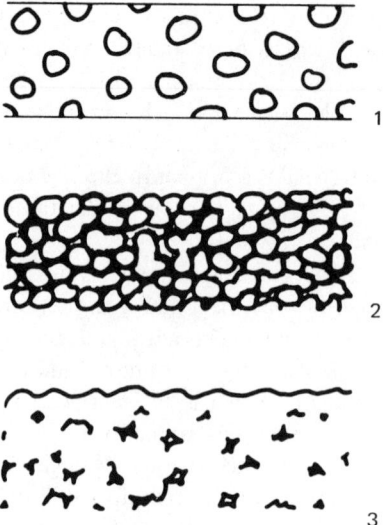

Figure D-1   Stages in the Drying of a Polymer Emulsion.

continuous coating of the polymer, has been the subject of considerable debate and study. The Division of Organic Coatings and Plastics Chemistry of the American Chemical Society awarded the Union Carbide Chemical Prize of 1965 to E. G. Bradford and J. W. Vanderhoff, of the Dow Chemical Company, for their electron microscopic studies that clearly demonstrated the gradual formation of the continuous film. Bradford and Vanderhoff state that a number of reasons can be given to explain why the polymer droplets flow together. Essentially, the particles of polymer flow together for much the same reasons that two drops of water on a table top tend to flow together when they touch. Flow can be facilitated by the presence of plasticizers,

TABLE D-1.   VARIATION OF RESIDUAL WATER CONTENT WITH DRYING TIME FOR A 67/33 STYRENE/BUTADIENE CO-POLYMER EMULSION (BRADFORD AND VANDERHOFF)

TIME	PER CENT WATER IN THE FILM	PER CENT OF TOTAL WATER EVAPORATED
0	50.06	–
2.5 hrs.	2.70	97.2
6.5 hrs.	1.99	98.0
1 day	1.07	98.9
2 days	0.63	99.4
5 days	0.28	99.7
7–11 days	(0.00 ± 0.05)	(100.0 ± 0.05)

which lowers the viscosity of the polymer. The particles can also be made to flow together more rapidly through the application of heat. A continuous organic coating is formed when stage 3 has fully been achieved, perhaps in a month or two. However, this depends on several factors, including the presence of plasticizers and the application of heat. The net result is that the toughness of a polymer-emulsion coating usually increases with time. The electron microphotographs produced by Bradford and Vanderhoff clearly showed the transition from a collection of particles touching one another to the final continuous film.

In the state represented by stage 3, the film usually can only be dissolved with organic solvents. In other words, when a polymer emulsion has thoroughly aged, it is usually no longer possible to disperse it effectively with water. At this stage, organic solvents must be used to dissolve the coatings, just as is necessary with any other film of thermoplastic resin. Under the circumstances, the solvent required to dissolve the final film represents another key property to be evaluated in choosing polymer emulsions.[7] (The Rohm & Haas Company's acrylic resin emulsions Rhoplex ® AC-33 or AC-55 are soluble in toluene and xylene.) Because the molecular weight of the polymer is rather high, such films often seem to go into solution more slowly than one would expect of a thermoplastic resin; they usually form a solution of considerable viscosity.

The formation of films from polymer emulsions depends upon the ability of the finely dispersed droplets of polymer to flow together once the water has evaporated. A characteristic temperature is observed, known as the minimum film formation temperature (MFT), below which the flowing together of the droplets does not take place in polymer emulsions and the emulsion does not form a satisfactory film. The MFT is practically independent of the particle size of the dispersed polymer but is related to surface tension and, particularly, to the stiffness of the polymer.[8]

The composition of co-polymers, of course, sharply influences the stiffness of the films. Data in Table D-2 illustrate the change in MFT observed as the composition of an acrylate/methacrylate co-polymer is varied. Values of MFT are also compared with the glass transition temperature (T_g) of the co-polymers.[9] This particular property of emulsions can become especially important to the artist or conservator if he must work outdoors or in unheated buildings where the temperature in certain seasons may be lower than the MFT.

A review of the properties of acrylic paints as used in the fine arts has appeared in *Resin Review*, 16, No. 3 (1966): 12-16, published by the Rohm & Haas Company. Among the many details mentioned, it is pointed out that antifoaming agents may be used profitably in emulsion paints. The

TABLE D-2. MINIMUM FILM FORMING TEMPERATURE (MFT) AND GLASS TRANSITION TEMPERATURE (T_g) OF POLYMER EMULSIONS AS A FUNCTION OF THE COMPOSITION OF CO-POLYMERS BASED ON METHYL METHACRYLATE (MMA) AND ETHYL ACRYLATE (EA) [KING AND NAIDUS]

PER CENT MMA/EA	MFT, °C.	T_g, °C.
100:0	109	105.8
80:20	78	77
65:35	54	–
60:40	47	–
55:45	39	–
50:50	32	38.9
20:80	-15	0.8
0:100	-46.5	-22

Nopco 1497V, NDW, or NXZ types are among those recommended (Nopco Chemical Company, 60 Park Place, Newark, New Jersey, 07102). The review also discussed at considerable length the use of dispersing agents, such as Rohm & Haas' Tamol 731 or Triton CF-10, for the effective dispersal of pigments in emulsion paints. To reduce the tendency of the paint system to mildew or mold, the addition of a preservative may also be advisable. Mercury compounds may be sensitive to hydrogen sulfide in the atmosphere and can produce a black or brown discoloration in the paint film. Because of this, agents such as Nicon P (Vetanicon Chemical and Drug Company, Inc.) and tributyl tin oxide at a concentration of about 1% are suggested instead. A number of thickening agents are listed. It is also pointed out that suppliers may occasionally recommend that artists use ethylene glycol or propylene glycol to slow down the rate of evaporation of the water.

Emulsion coatings are noted for their ability to transmit water vapor. They allow water vapor to pass in and out of wood and masonry, reducing the tendency of the films to blister. High permeability to water vapor can present problems; for example, coatings on masonry often exhibit efflorescence. Bondy has suggested that this can be minimized by the selection of hydrophobic polymers and by reducing the water-soluble components in the formulation.[10] Alkaline surfaces can also cause the deterioration of coatings which are not sufficiently resistant to hydrolysis.

The remaining major variable to consider is particle size. The particles of the dispersed polymer in the usual polymer emulsion vary in size from about 0.1 to 10 μ, with the range from 0.5 to 3 μ perhaps most commonly encountered (25.4 μ = 0.001").

Modern pigments cover a range of sizes generally lower than 5 μ; many artist's pigments in the past perhaps fell largely in the range from 5 to 10 μ.

On the basis of the relative size of pigments and polymer emulsion particles, it can be seen that emulsions of low particle size are able to bind pigments most efficiently. Martins reports data indicating that the critical pigment volume concentration (CPVC) of emulsion paints may be about 30% when the particle size is 5 to 10 μ, but can be as high as 50% when the emulsion particles are only 0.1 to 0.2 μ.[11] (The CPVC occurs when the voids in a paint are either just filled by the vehicle or at a minimum.) When it is desired to minimize the darkening of chalky pigments by a consolidant, the larger particle-sized emulsions may not fill the voids between the pigment particles as well as a more finely dispersed emulsion would.

Bondy has published a diagram similar to Figure D-2 in order to draw attention to properties that may be expected to be influenced by the size of the dispersed polymer particles.[10] In the absence of intentionally added thickeners, it can be seen that the viscosity of a given concentration of emulsion will increase as the particle size is reduced. This is due largely to the individual particles being closer together. The gloss of emulsion paints is also usually improved by lowering the particle size; usually improvements are also noted in water resistance, the stability of the emulsion, and the tensile strength of the resulting film.

The high surface tension of water-based coatings usually reduces their tendency to penetrate porous surfaces. Increasing the particle size will further lower the ability of the polymer emulsions to penetrate, improving what is known as "hold-out," the tendency of the coating to concentrate on the surface of the paper, textiles, fiberboard, wood, concrete block, etc. One manufacturer, for example, offers a series of wood adhesives formulated at

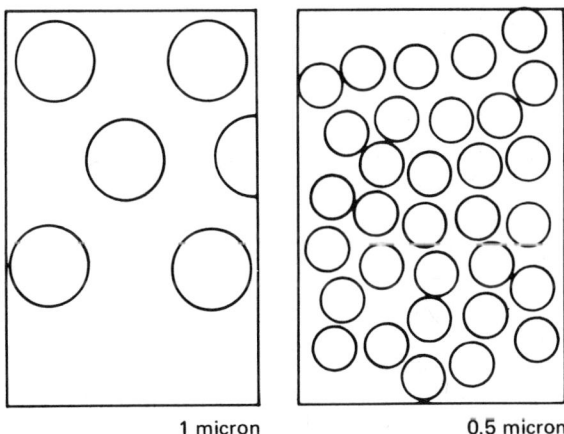

Figure D-2 Two Polymer Emulsions at the Same Concentration by Volume.

10-μ particle size, undoubtedly to aid the adhesive in remaining on the surface rather than penetrating excessively into the wood.

This brief description of the properties of polymer emulsions and how they dry to form continuous polymer films, has pointed out key qualities to be kept in mind. In choosing emulsions, one of the first properties to test is the acidity or alkalinity of the polymer emulsion; types which are practically neutral are available. The strength that the film can develop will depend ultimately on the molecular weight of the polymer in the emulsion, but the speed at which the final strength is attained depends on the ease with which the polymer particles flow together.[12] A third key property to be kept in mind is that, once the polymer particles have flowed together (stage 3), organic solvents are usually needed to dissolve the coating. At this stage, polymer emulsion coatings vary in the "strength" of solvent required to remove them. If the coatings are to be used at low temperatures, the MFT must be taken into account. The particle size of an emulsion can significantly influence its ability to penetrate porous materials. Finally, the presence of volatile plasticizers and the tendency to cross-link with age should be considered.

In selecting an emulsion from among the wide variety offered, the following checklist may be useful:

> pH
> chemical composition
> hardness of the final film (a low minimum film-forming temperature, MFT, or glass-transition temperature, T_g, generally indicates low hardness)
> water resistance
> solubility of the final polymer film in organic solvents
> presence and type of plasticizer
> molecular weight of polymer and strength of final film
> solids content
> viscosity and type of thickener if present
> ionic charge on particles and stability of emulsion towards salts and freeze-thaw cycles
> particle size
> need for fungicide

REFERENCES AND NOTES

1. Reprinted with slight changes from R. L. Feller, "Polymer Emulsions," *Bulletin of the American Group-IIC* 6, No. 2 (1966): 24-28; 7, No. 1 (1967): 18-19; 9, No. 2 (1969): 15-17.
2. See also (a) F. Denninger, "Künststoffdispersionen in Malerei und Restaurierung," ("Synthetic Resin Dispersions in Painting and Restoration")

Maltechnik 4 (1960): 97-105 [in German]; *IIC Abstracts*, 3495; (b) W. G. Vannoy, "Emulsion Paints on Wood," *Official Digest, Federation of Paint and Varnish Production Clubs* 33 (1961): 807-29; (c) R. Wihr, "Bergung und Konservierung römischer Wandmalereien," ("Recovery and Conservation of Roman Mural Paintings"), *Ergänzungsbände des Berliner Jahrbuchs für Vor- und Frühgeschichte* 1 (1964): 98-101 [in German]; *IIC Abstracts*, 5325; (d) C. R. Martins, *Emulsion and Water-soluble Paints and Coatings* (New York; Reinhold, 1964).

3. The importance of thickening agents has been pointed out in the three paragraphs on the subject of emulsions which appear in *Synthetic Materials Used in the Conservation of Cultural Property* (Rome Centre, 1963), p. 13.
4. C. J. Piez, "Some Library Adhesives—A Laboratory Evaluation of PVA's," *Am. Library Assoc. Bulletin* 56 (1962): 838-43; *IIC Abstracts*, 3835.
5. For additional literature: Address the Special Products Department, Rohm & Haas Company, Philadelphia, Pennsylvania 19105.
6. Rohm & Haas Company, Special Products Department, Bulletin on *Emulsion Polymerization of Acrylic Monomers* (Philadelphia, 1965).
7. Cotton-duck and linen canvases in America are now being preprimed with polymer emulsions. The Grumbacher Company says that they have been preparing canvases in this way for over fifteen years. Mr. H. Levison, former president of Permanent Pigments, pointed out that a further change has occurred: their canvases prepared with methacrylate polymer emulsions are no longer glue-sized before priming. Conservators should keep in mind that the present types of polymer-emulsion grounds are soluble in toluene and, occasionally, in slightly "milder" solvents.
8. J. G. Brodnyan and T. Konen, "Experimental Study of the Mechanism of Film Formation,": *J. Appl. Polymer Sci.* 8 (1964): 687-97.
9. A. J. King and H. Naidus, "The Relationship Between Emulsion Freeze-Thaw Stability and Polymer Glass Transition Temperature—1. A Study of the Polymers and Co-polymers of Methyl Methacrylate and Ethyl Acrylate," *Polymer Preprints, American Chemical Society, Papers Given at the New York Meeting* 7, No. 2 (September 1966): 860-69.
10. C. Bondy, "Binder Design and Performance in Emulsion Paints," *J. Oil & Colour Chemists' Assoc.* 51 (1968): 409-27.
11. C. R. Martins, *Emulsion and Water-soluble Paints and Coatings* (New York: Reinhold, 1964).
12. Coating a polymer emulsion on a glass plate, allowing it to dry, and testing its toughness with a knife, can sometimes quickly reveal differences in the film properties of the coatings sold by various manufacturers.

Appendix E

GRADES OF POLY(VINYL ACETATE) RESINS WITH RESPECT TO THEIR VISCOSITY IN SOLUTION

Robert L. Feller

Table E-1 will be of assistance in comparing different proprietary poly(vinyl acetate) resins with respect to their viscosity in solution. The principal data were first presented in a paper by R. L. Feller on the "Identification and Analysis of Resins and Spirit Varnishes" given at a seminar held at the Museum of Fine Arts, Boston, 15-18 September 1958, and later published with the conference papers (*Application of Science in Examination of Works of Art*, Boston, 1959). The original table was based in part upon data reported by Shaw, as well as upon measurements made at Mellon Institute, and information obtained through the kind assistance of Dr. A. E. Werner, Research Laboratory of the British Museum, London, Dr. H. Laemmel, Union Carbide Europa, Geneva, and Dr. S. Augusti, Soprintendenza alle Gallerie ed Opere d'Arte, Naples.[1] Since the first edition, a more extensive table has appeared in a booklet published by the Rome Centre, later brought up-to-date and reprinted in an appendix to a recent UNESCO publication.[2,3]

The kind cooperation of the manufacturers is gratefully acknowledged. Without their assistance regarding specific questions, such as the relationship between former designations and present ones, the preparation of such a table would have been exceedingly difficult.

The relationship between the various products in the table is shown only with respect to their viscosity in solution, and is for many reasons only an

TABLE E-1. POLY(VINYL ACETATE) RESINS FROM VARIOUS MANUFACTURERS

Characterizing Measurements							
Viscosity (86.1 g./L)	(a)	1.5	2.5	5		15	21
Viscosity Grade	(b)		9	40		80	167
Intrinsic Viscosity	(c)	0.11	0.15	0.39		0.56	0.69
Approximate DP	(d)	130	210	520		1040	1500
Approximate T_g, °C	(d)	16	17	21		24	26
Trade Designations							
Gelva	(e)	1.5	2.5	4	7	15	25
Union Carbide Vinyl Resins	(f)	AYAC	AYAB	AYAA		AYAF	AYAT
Mowilith	(g)		20	30	40	50	60
Vinalak	(h)		5249		5254	5268	
Vinavil	(i)		K25	K40		K50	
Rhodopas	(j)	BB	B	M			H
Vinnapas	(k)	B1.5	B5	B17	B60	UW1	UW4

(a) Viscosity in centipoises, benzene solution containing 86.1 g. of resin per 1000 cc., of solution at 20°C., given in the commercial technical literature.

(b) Viscosity in centipoises of a solution in toluene at 20% solids by weight; 70°F. (21.1°C.); data of S. Raynolds, Mellon Institute.

(c) Determined in cyclohexanone at 20°C.: data from technical literature of Union Carbide Corporation.

(d) Values of T_g and DP reported in *A Handbook of Adhesives*, I. Skeist, (New York: Reinhold, 1962); and also in *The Conservation of Cultural Property*, UNESCO, 1968, 312.

(e) Product of Shawinigan Products Corporation, Shawinigan Falls, P. Q. Canada. Shawinigan Products Corporation has used the measurement (a) above as their designation of grade.

(f) Product of Bakelite Division, Union Carbide Corporation, New York, New York, 10017, U.S.A.

(g) Product of Farbwerke Hoechst AG, Frankfurt (M)–Hoechst, Federal Republic of Germany.

(h) Product of Vinyl Products, Ltd., Butter Hill, Carshalton, Surrey, England.

(i) Products of Societa Rhodiatoce, Gruppo Montecatini, via F. Turati, 18 Milan, Italy.

(j) Product of Société des Usines Chimiques Rhone-Poulenc, 21 rue Jean-Goujon, Paris 8, France.

(k) Product of Wacker-Chemie GmbH, Prinzregentenstrasse 22 Munich, Federal Republic of Germany.

approximate arrangement. Each investigator must determine for himself the viscosity grade or intrinsic viscosity of the polymer which he is employing if a precise comparison is desired.

Shaw reported data on Vinnapas poly(vinyl acetate) resins from Alexander Wacker Gesellschaft, Munich. This company offers a range of viscosity grades between 9 and 80 which should be particularly useful to conservators.[1]

To permit further comparisons: the viscosity grades of Bedacryl 122X (Imperial Chemical Industries) and Elvacite ® 2044 (du Pont) poly(n-butyl methacrylate) are about 48 cp ; the viscosity grades of dammar, mastic, and resin AW-2 fall between 1.2 and 1.8 cp. (see Table 5-1).[4,5,6]

Values of average molecular weights or degrees of polymerization in such tables are only approximations. Care must be taken in noting and comparing molecular weights because the number-average molecular weight, determined by osmometry, is usually lower than the viscosity-determined value and that, in turn, is lower than the weight-average molecular weight, determined by light scattering or the ultra-centrifuge.

Indication of the molecular weights of these polymers may be obtained from the following: Shaw reported molecular weights of 4300 for Gelva 2.5 and 25,000 for Gelva 15, based on the viscosity-molecular weight relationship of Staudinger.[1] Sutherland and Funt have reported the molecular weight of Gelva 60 as 180,000.[7] Blom has given an interesting curve of the molecular weight distributions in Gelva 2.5, 15, and 45, in terms of chain length; the average values resemble those reported by Shaw.[8] Values for the intrinsic viscosities and the viscosity-average molecular weights of Gelva 6, 15 and 25, have been compared with number-average molecular weights, ranging from 52,000 to 114,000, by Wagner and others.[9,10] Skeist has reported the weight-average molecular weights.[11]

The viscosity-average molecular weight of vinyl resin AYAC has been reported to be about 3300 by Hecht, Coles, and Keeler.[12] Schildknecht has reported the molecular weight of Mowilith 30 as 37,000.[13] The weight-average molecular weights of Mowilith 20, 30, and 50, determined from light scattering, have been reported by the manufacturer to be 35,000, 110,000 and 260,000, respectively; it is expected that these values will be higher than number- or viscosity-average molecular weights.[14] The relationship of the intrinsic viscosity of poly(vinyl acetate) in various solvents to the number-average molecular weight has been reported by Moore and Murphy.[15]

The earliest variety of poly(vinyl acetate) from the Bakelite Division of the Union Carbide Corporation was known as "Vinylite A,"[1] similar to the present AYAF. This was the viscosity grade first used by the Fogg Art Museum.[16,17]

REFERENCES AND NOTES

1. T. P. G. Shaw, "The Properties of Some Polyvinyl Resins as Lacquer Resins," *Official Digest, Federation of Paint and Varnish Production Clubs* No. 162 (1937): 9-18.
2. *Synthetic Materials Used in the Conservation of Cultural Property* (Rome: International Centre for the Study of the Preservation and the Restoration of Cultural Property, 256 Via Cavour, Rome, 1963).
3. "Synthetic Materials Used in the Conservation of Cultural Property," *The Conservation of Cultural Property* (Paris: UNESCO, 1968), Appendix, pp. 303-38.

4. R. L. Feller, "Factors Affecting the Appearance of Picture Varnish," *Science* 125 (1957): 1143-44.
5. R. L. Feller, "Dammar and Mastic Varnishes—Hardness, Brittleness, and Change in Weight Upon Drying," *Studies in Conservation* 3 (1958): 162-74.
6. R. L. Feller, "New Solvent-type Varnishes," in *Recent Advances in Conservation* (London: Butterworths, 1963), pp. 171-75.
7. T. H. Sutherland and B. L. Funt, "Dielectric Relaxation in a Series of Polyvinyl Acetyls," *J. Polymer Sci.* 11 (1953): 177-86.
8. A. V. Blom, "Synthetic Film-forming Materials: Vinyl and Allied Polymers," *Organic Coatings as Theory and Practice* (New York: Elsevier, 1949), p. 101.
9. R. H. Wagner, "Intrinsic Viscosities and Molecular Weights of Polyvinyl Acetates," *J. Polymer Sci.* 2 (1947): 21-35.
10. R. E. Robertson, R. McIntosh, and W. E. Grummitt, "The Influence of Membrane Preparation on the Osmotic Pressure of Polyvinyl Acetate in Acetone," *Canadian J. Res.* B24 (1946): 150-66.
11. I. Skeist (ed.), *A Handbook of Adhesives* (New York: Reinhold, 1962).
12. Society of Chemical Industry, "The Development of a Characteristic Constant for the Determination of the Adhesive Properties of High Polymers," *Adhesion and Adhesives: Fundamentals and Practice* (New York: Wiley, 1954), p. 61.
13. C. E. Schildknecht, "Vinyl Acetate Polymer Aqueous Dispersions ('Emulsions')," *Vinyl and Related Polymers* (New York: Wiley, 1952), p. 333.
14. Farbwerke Hoechst, Frankfurt (M)—Hoechst (Private communication).
15. W. R. Moore and M. Murphy, "Viscosities of Dilute Solutions of Polyvinyl Acetate," *J. Polymer Sci.* 56 (1962): 519-32.
16. G. L. Stout and R. J. Gettens, "Transport des Fresques Orientales sur de Nouveaux Supports," *Mouseion* 17 (1932): 107-12.
17. R. J. Gettens, "Polymerized Vinyl Acetate and Related Compounds in the Restoration of Objects of Art," *Technical Studies in the Field of the Fine Arts* 4 (1935): 15-27.

Appendix F

PHYSICAL, CHEMICAL, AND TOXICOLOGICAL PROPERTIES OF SOLVENTS AND LIQUIDS FOR CONSERVATION OF PAINTINGS AND WORKS OF ART

Nathan Stolow

NOTES

Evaporation Rate: Evaporation rate data obtained from "Handbook of Organic Industrial Solvents," *National Association of Mutual Casualty Companies* (Chicago, 1958) 72 pp. These data are originally taken from Wilson, L.D., "Evaporation rate of solvents and an improved method for their determination," *Paint, Oil and Chemical Review,* 118, No. 24 (1955): p. 6. The evaporation rates given are relative to diethyl ether, of value 1.

Fire Hazard: VF—very flammable, vapors can be explosive; F—flammable; SL—flammable at high temperatures; NIL—not flammable. Solvents with low flash points, i.e., VF, are acetone, benzene, cyclohexane, ethyl acetate, methyl ethyl ketone, tetrahydrofuran, petroleum ether, diethyl ether.

Toxicity: The toxicity numbers are quoted in Sax, "*Dangerous Properties of Industrial Materials,*" 2nd edition (New York: Reinhold, 1963). 1343 pp. 0 = No apparent toxicity, 1 = slight, 2 = moderate, 3 = high, and U = unknown toxicity (but may be very toxic nevertheless).
Also refer to "*Toxicity and Metabolism of Industrial Solvents,*" Ethel Browning (New York: Elsevier, 1965). 739 pp.

M.A.C.: Maximum allowable concentration, refers to the threshold toxic concentration of the substance measured in *parts of toxic vapor per million parts of air.* Where this is not known the indication is U. This does not necessarily mean that the substance is safe, but that it has not been adequately studied as to physiological action. In all cases ventilation is recommended. M.A.C. data from Sax (loc. cit.).

Peroxides in ethers, etc: Ethers and similar solvents suspected of containing dangerous peroxides may be rendered safe by pre-treatment with strong ferrous sulfate solution made slightly acid with sodium bisulfate.

NAME, SYNONYM	FORMULA	BOILING POINT °C.	DENSITY	SOLUBILITY PARAMETER	EVAPORATION RATE	FIRE HAZARD	TOXICITY	M.A.C.	REMARKS
Acetone	CH_3COCH_3	56.5	0.797	10.0	1.9	VF	2	1000	
n-Amyl acetate (banana oil)	$CH_3COO(CH_2)_4CH_3$	148.0	0.879	8.5	11.6	VF	2	200	
n-Amyl alcohol	$CH_3(CH_2)_3CH_2OH$	138.0	0.817	10.9	38.1	F	2	100	
Benzene	C_6H_6	80.1	0.879	9.2	2.8	VF	3	25	
n-Butyl acetate	$CH_3COO(CH_2)_3CH_3$	126.0	0.88	8.5	7.8	F	2	200	
n-Butyl alcohol	$CH_3(CH_2)_2CH_2OH$	117.5	0.810	11.4	19.6	VF	2	100	
Isobutyl alcohol	$(CH_3)_2CHCH_2OH$	108.0	0.805	11.1	16.3	VF	3	U	
n-Butylamine (1-aminobutane)	$C_4H_9NH_2$	77.8	0.740	—	—	F	2	5	
Butyl cellosolve (ethylene glycol monobutyl ether)	$C_4H_9OCH_2CH_2OH$	171.0	0.903	8.9	85.0	F	2	50	
Carbon tetrachloride	CCl_4	76.8	1.595	8.6	2.6	NIL	3	25	
Cellosolve (ethylene glycol monoethyl ether)	$C_2H_5OCH_2CH_2OH$	135.1	0.936	9.9	28.1	F	1	200	
Cellosolve acetate (2-ethoxy acetate; ethylene glycol monoethyl ether acetate)	$CH_3COO(CH_2)_2OCH_2CH_3$	156.4	0.975	8.7	32.4	F	2	U	
Chloroform	$CHCl_3$	61.3	1.50	9.3	2.2	NIL	3	100	
Cyclohexane	C_6H_{12}	80.7	0.779	8.2	2.6	VF	2	400	
Cyclohexanol (hexalin)	$H_2COH(CH_2)_4CH_2$	161.5	0.945	11.4	150.0	VF	1	100	
Cyclohexanone	$\overline{CO(CH_2)_4CH_2}$	155.6	0.948	9.9	22.2	F	1	100	
Diacetone alcohol	$CH_3COCH_2C(CH_3)_2OH$	168.0	0.931	9.2	60.0	F	1	50	Yellows on standing
Diethyl benzene	$(C_2H_5)_2C_6H_4$	184.9	0.868	8.7	—	F	U	U	
Diethyl ether (ether)	$C_2H_5OC_2H_5$	34.6	0.714	7.4	1.0	VF	2	400	May explode on burning, evaporation leaves explosive peroxides

Name	Formula	BP	Density						Notes
Dimethyl formamide	(CH$_3$)$_2$NCHO	152.8	0.945	12.1	—	F	3	U	
1,4 Dioxane (diethylene oxide; diethylene dioxide)	OCH$_2$CH$_2$OCH$_2$CH$_2$	101.3	1.035	10.0	5.8	VF	2	100	Forms dangerous peroxides on exposure to air—not to be distilled unless peroxides are destroyed
Dipentene	C$_{10}$H$_{16}$	175.0	0.86	8.5	—	F	2	U	
Di-isopropyl benzene	[(CH$_3$)$_2$CH]$_2$C$_6$H$_4$	205.0	0.863	8.7	—	F	U	U	
Ethanol (ethyl alcohol)	CH$_3$CH$_2$OH	78.3	0.789	12.7	7.0	VF	1	1000	Oxidizes on storage
Ethyl acetate	CH$_3$COOC$_2$H$_5$	77.2	0.895	9.1	2.7	VF	2	400	
Ethyl benzene	C$_2$H$_5$C$_6$H$_5$	136.2	0.870	8.8	9.4	VF	2	200	
Ethylene dichloride (1,2, dichloroethane)	CH$_2$ClCH$_2$Cl	83.5	1.26	9.8	3.3	VF	3	100	
Glycerol (glycerine)	CH$_2$OHCHOHCH$_2$OH	290.0	1.26	16.5	hygrosc.	SL	1	NIL	Absorbs H$_2$O on storage
n-Heptane	CH$_3$(CH$_2$)$_5$CH$_3$	98.5	0.684	7.4	2.7	VF	1	500	
Kerosene	Petroleum hydrocarbons, 10-16C	175.-325.	0.90	7.6	—	F	1	U	
Methanol (methyl alcohol) (wood alcohol)	CH$_3$OH	64.3	0.791	14.5	5.2	VF	2	200	
Methyl acetate	CH$_3$COOCH$_3$	57.2	0.935	9.6	2.2	VF	2	200	
Methyl isobutyl ketone (hexone)	CH$_3$COCH$_2$CH(CH$_3$)$_2$	115.1	0.802	8.4	5.6	VF	2	100	
Methyl cellosolve	CH$_3$OCH$_2$CH$_2$OH	124.5	0.966	10.8	21.1	F	2	25	
Methylene chloride	CH$_2$Cl$_2$	40.0	1.34	9.7	1.8	NIL	3	500	
Methyl ethyl ketone (2-butanone)	CH$_3$COCH$_2$CH$_3$	79.6	0.806	9.3	2.7	VF	1	250	
Methyl n-propyl ketone	CH$_3$CO(CH$_2$)$_2$CH$_3$	102.3	0.812	8.7	—	VF	2	200	
Morpholine	OCH$_2$CH$_2$NH-CH$_2$CH$_2$	128.9	0.998	—	—	F	2	U	
Isooctane (2,2,4-trimethylpentane)	(CH$_3$)$_2$CHCH$_2$C(CH$_3$)$_3$	99.2	0.703	7.3	5.9	VF	1	500	
n-Octanol (caprylic alcohol)	CH$_3$(CH$_2$)$_6$CH$_2$OH	195.0	0.827	10.5	—	F	U	U	
Petroleum ether (petroleum naphtha, ligroin)	Low-boiling petroleum hydrocarbons (1-5% aromatics)	40.70	0.63	7.0	—	VF	1	500	

NAME, SYNONYM	FORMULA	BOILING POINT °C	DENSITY	SOLUBILITY PARAMETER	EVAPORATION RATE	FIRE HAZARD	TOXICITY	M.A.C.	REMARKS
n-Propyl acetate	$CH_3COOC_3H_7$	101.6	0.89	8.8	4.8	VF	2	200	
Isopropyl alcohol	$(CH_3)_2CH-OH$	82.3	0.785	11.5	7.7	VF	2	400	
n-Propyl alcohol	$CH_3CH_2CH_2OH$	97.2	0.804	11.9	7.8	VF	1	400	
Pyridine	$NCHCHCHCHCH$	115.3	0.982	10.7	8.2	VF	3	10	
Shell cyclosol 53 (aromatic solvent)	Petroleum distillate with KBV 93.7	150-160	0.873	8.5	–	F	U	U	
Shell sol 715 (72) (odorless)	Petroleum distillate with KBV 27.5	185-200	0.758	7.0	–	F	1	U	
Tetrahydrofuran (cyclotetra methylene oxide)	$O(CH_2)_3CH_2$	65.4	0.888	9.9	2.0	VF	U	200	Forms dangerous peroxides on exposure to air – not to be distilled unless peroxides are destroyed
Turpentine	Mainly $C_{10}H_{16}$	154-170	0.85	8.5	375.0	F	2	100	Oxidizes on storage
Tetrachloroethylene (perchloroethylene)	CCl_2CCl_2	121.0	1.63	9.4	6.6	NIL	3	200	Decomposes on storage, usually stabilizing agent added
Trichlorethylene	$CHClCCl_2$	86.9	1.46	9.3	3.1	NIL	2	200	
1,1,2 Trichloroethane	$C_2H_3Cl_3$	113.7	1.443	8.5	12.6	NIL	2	U	
Triethanolamine	$(HOCH_2CH_2)_3N$	360.0	1.126	–	hygrosc.	slight	1	1	
Toluene	$C_6H_5CH_3$	110.6	0.872	8.9	4.5	VF	2	200	
Varsol DX 3139	Petroleum distillate with KBV 39	150-170	0.777	7.0	–	VF	1	U	
V.M. & P. Naphtha (76° benzine)	Petroleum distillate with KBV 46	100-140	0.75	7.6	7.1	VF	1	U	
White Spirit B.S. 245-1956	Petroleum distillate British, approx. 16-20% aromatics	155-210	0.80	7.6	–	VF	U	U	
Xylene	$(CH_3)_2C_6H_4$ mixture o, m, p-xylenes	138-145	0.87	8.8	9.2	VF	1	200	

INDEX

Certain solvents that appear only in Appendix F are not listed in this index

Abrasion, 146, 152, 181
Absorbers. *See* Ultra-violet absorbers
Absorption spectra, of various materials, 148. *See also* Color; Infra-red spectra; Ultra-violet radiation
Accelerated aging test, 160, 162, 163, 182, 208. *See also* Daylight; Exposure; Fadeometer; Ultra-violet radiation
Acetic acid, 17, 18, 32
Acetone, 4, 18, 21, 89, 174, 175, 176, 212
  diffusion rate and coefficients of, 33, 99, 102-5 *passim*
  evaporation time of, 26, 110, 231
  leached components analyzed, 70, 80, 83, 88
  leaching and swelling by, 55-59 *passim*, 61-69 *passim*, 109-10
  in removal: of cross-linked polymer, 121, 153, 154, 163, 205, 206; of dammar varnish, 49, 106, 121, 165; of oil films, 109-11
  solubility parameter of, and swelling by, 91-95 *passim*
  toxicity and safety of, 39, 231
Acetylenic hydrocarbons, 14
Acidic groups formed upon oxidation, 121, 161
Acidity, of polymer emulsions, 219, 224
Acids, organic, 16, 17, 18, 32. *See also* Dicarboxylic acids
Acrylic. *See* Polymers, acrylic *and* methacrylate

Acryloid®, polymers, 34, 37, 122, 124, 125, 127, 129, 132, 161, 200, 212
Activation of deterioration, 6, 156
Addition polymer, 126
Adhesive, 127, 150, 223
Adipic acid, 53
Age of varnish, periods in, 4-6. *See also* Accelerated aging test; Deterioration; Re-forming; Removal
Alcohols, 14-17 *passim*, 47, 48, 90. *See also* Ethanol; n-Propanol; etc.
  used in re-forming, 180
  diffusion rate of, 33
Aliphatic compounds, 36. *See also* Hydrocarbons; Paraffin hydrocarbons
Alkalinity of polymer emulsions. *See* Acidity
Alkyd, 28, 148
Alkyl benzenes, 36-38
Aluminum foil, as support for films, 123, 124, 127, 128, 138, 157, 161, 203-6 *passim*, 211-17 *passim*
Alvar. *See* Polymers: poly (vinyl acetal)
Ammonia and amines, 21, 126
Ammonium sulfate, in bloom, 155
Amorphous material, 27, 119, 120, 151, 152, 154
n-Amyl alcohol, 89, 91, 92, 102, 103, 104, 231
Analytical grade solvent, 25
Aniline, 21, 32
Antifoaming agents, 221, 222

236 INDEX

Antioxidant, 159, 160. See also Inhibition; Ultra-violet absorbers
Appearance of varnish. See varnish
Aromatic hydrocarbons, 10-12, 34, 36-38, 122, 139
  character of, 14, 36-37
  swelling behavior of, 90
Atom and atomic weight, 7
Autoxidation, 52, 160
Average degree of polymerization, 126, 163
AW-2 resin, 34, 122, 124, 126, 128, 129, 132
  appearance of, 141, 142, 154
  resistance to deterioration of, 133, 160
AYAA, AYAB, AYAC, AYAF, AYAT. See Polymers: poly (vinyl acetate), Union Carbide products
Azelaic acid, 53

Badische Anilin und Soda Fabrik, 122
Bakelite® products, 122, 127, 129, 227. See also Polymers: Poly(vinyl acetate), Union Carbide products; poly(vinyl chloride-acetate)
Baking films, to drive off solvent, 5, 124
Balanced formulation of solvent, 139
BASF. See Badische Anilin und Soda Fabrik
Bedacryl 122X, 227
Beeswax, solubility of, 41
Bending test. See Brittleness; Mandrel-bend test
Benzene, 10, 11, 12, 14, 22, 32, 36, 38, 89, 195
  diffusion rate and coefficients of, 99, 102-4
  evaporation time of, 4, 26, 39, 231
  leaching and swelling by, 107, 110
  solubility parameter of and swelling by, 91-94 passim
  toxicity and safety of, 39, 40, 231
Benzine, 12, 13
Benzophenones, 149, 160, 199, 200
Bleaching of varnish color, 124
Bloom
  occurrence of, 5, 124, 152, 155
  type of gloss, 139, 140
Blushing, 5

Boiling point. See also specific compounds
  of alkyl benzenes, 37, 38
  as indication of volatility, 26, 126
  range, significance of, 8, 12, 13, 14, 22
Bonds, 8, 10, 11, 14, 188, 189. See also Hydrogen bonding
  in cross-linking, 203, 216
Boron trifluoride-methanol, esterification with, 75
Brittleness. See also Chain breaking; Cracks; Elongation-at-break
  influenced by retained solvent, 4, 5, 128, 139
  removal facilitated by, 164, 206
  of various polymers and resins, 123-28 passim
Brittle point, 4, 119
n-Butane, 10
Butyl acetate, toxicity of, 40, 231
n-Butyl alcohol, in Kauri-Butanol test, 34
t-Butylbenzene, influence on brittleness by, 138
Butyric acid, 18

Canvas, 150, 225
Carbon-arc. See Fadeometer
Carbon dioxide lost upon aging, 161
Carbon-14. See Radioactive ethanol
Carbon tetrachloride, 21, 36, 39, 89, 231
  diffusion rate and coefficients of, 96-98, 101-2
  leached components analyzed, 71
  leaching and swelling by, 65, 110
  solubility parameter of and swelling by, 91-92
Carboxylic acids, 16, 17, 18, 32. See also Acids, organic; Dicarboxylic acids
Cellosolve, 15, 89
  diffusion rate and coefficients of, 103
  evaporation time of, 231
  leaching and swelling by, 57, 64-65, 67, 69, 109, 110
  solubility parameter of and swelling by, 91-92
  toxicity and safety of, 231
Cellosolve acetate, 21, 137, 141, 178, 231
Cellulose acetate, 125

INDEX 237

Cellulose acetate-butyrate, 153, 172
 chain breaking of, 155, 156, 162
Cellulose acetobutyrate. *See* Cellulose acetate-butyrate
Cellulose nitrate, 5, 28, 29, 125, 129
Ceresin wax, 9
Chain breaking
 effect of temperature on, 196, 215
 effects of, 53, 155, 156, 162, 205
 ratio to cross-linking. *See* Cross-linking
Chalking, 154, 223
Checklist of polymer emulsion properties, 224
Chios, source of mastic, 120
Chlorinated hydrocarbons, 36, 40, 138, 174, 175, 176
 swelling behavior of, 90
 toxicity of, 40, 231. *See also specific names*
Chlorinated rubber, classification test for, 127
Chlorine
 as oxidizing agent, 155
 in polymers, 6, 133. *See also* Polymers: poly(vinyl chloride); etc.
Chloroform, 21, 36, 39, 89, 231
 evaporation time of, 110, 231
 leached components analyzed, 70, 79–81, 87
 leaching and swelling by, 65, 109, 110
 solubility parameter of, 91, 92
Chromatography, paper. *See also* Gas Chromatography
 in study of leached components, 72
 test for cross-linking, 158, 197–200 *passim*, 204, 207, 209
Classification
 of natural resins, 120
 of plasticizers, 130
 of polymers, 130, 131, 188
 of solvents, 25, 32, 40
 of varnishes, 3
Cleaned Pictures Exhibition, London, 50
Cleaning (pictures), 48–50, 105, 144, 179, 181, 189. *See also* Removal
Cloud point, 33–35 *passim*, 40
Coating, 146. *See also* Film; Natural resins; Paint; Polymers; Varnish

Cohesive energy density, 27, 90, 189. *See also* Solubility, parameter
Colloidal dispersion, 24
Colophony (rosin), discoloration of, 121
Color, of varnish, 5, 6, 124, 139, 142–45 *passim*, 149, 152, 153, 155, 173
Condensation polymers, 18, 126
Conservation, introduction of polymers in, 125, 127, 129, 130, 164, 228
Conservators' practices in formulating varnishes, 125, 127, 153
Constituents of dammar, mastic, 121
Constitution (of U.S.), protection of, 149
Contrast gloss, 139, 140
Control of appearance, 139–41
Copaiva balsam, 179
Copal, 172, 173, 175, 176, 184, 186, 187
Co-polymers. *See also* Acryloid®; Rhoplex® AC = 33; Rhoplex® AC = 55; *and co-polymers of specific chemical composition*
 of methyl methacrylate/ethyl acrylate, 132, 158, 161, 200, 222
Correlation of fadeometer vs. gallery exposure, 196
Cotton swabs. *See* Swabs
Cracks, in varnish, 147, 152, 154, 161, 173. *See also* Brittleness
Cross-link density, 162, 163, 205, 206
Cross-linking
 with chain breaking, 159, 162, 163, 205, 212–17 *passim*
 effect on swelling, 163, 202–7 *passim*
 effect of temperature on, 162, 196, 215
 effect of tertiary hydrogen on, 157, 158, 166
 effect of ultra-violet radiation on, 162, 164
 under gallery conditions: estimated, 158, 162, 164, 196, 200; observed, 197, 200, 207
 gel dose in, 162, 163, 205, 213, 215
 by heat, 156, 195
 inhibition of, 158, 159, 160, 199, 200, 203
 loss of solubility through, 155–60 *passim*, 162–64, 202–6 *passim*, 211–17 *passim*
 non-swelling stage in, 163, 164, 202, 206, 207

Cross-linking (*continued*)
  of polymer emulsions, 224
  polymers most sensitive to, 157, 158
  polymers, resistant to, 157, 158, 160, 196, 200
  ratio to chain breaking, determination of, 215
  test for, 164, 196, 197, 198, 204, 209, 213-17 *passim*
  of thermosetting polymers, 126
Crystallization, 154
Cyclohexane, 9, 10, 11, 14, 89, 91, 92, 121, 157, 163, 197, 200, 203-6 *passim*, 231
Cyclohexanol, 91, 231
Cyclohexanone, 91, 92, 94, 110, 231
Cycloparaffins, 12. *See also*, Cyclohexane; Methyl cyclohexane; Naptha

Damage per footcandle, 149, 153
Dammar resin, 120, 121
  brittleness of, 123, 124, 127, 128
  effect of light on, 121, 124, 160, 182
  hardness of, 122, 130, 132
Dammar varnish, 120, 121, 123, 151, 172
  appearance of, 124, 141-45 *passim*, 150, 154, 155, 173
  brittleness of, 128, 137, 138, 148
  effects of age and re-forming on, 175, 176, 183-87 *passim*
  loss of radioactive ethanol from, 186-87
  oxidation of, 121, 137, 155, 160, 161, 217
  reaction with zinc white, 154
  removal of, 47, 48, 121, 124, 165, 209, 210
  solvent retained by, 5, 137, 138, 182, 184, 187
  with stand oil, 123, 209
  transmission of light by, 143, 148, 149, 150
  viscosity of, 121-23, 139-42 *passim*
Dammar wax, 120, 121
Daylight, exposure to, 149, 158, 159, 164, 172, 174, 178, 196, 197, 200, 207, 209
Declaration of Independence, protection of, 149

Degradation. *See also* Weight loss
  volatile products of, 155, 161, 164, 195, 206, 212
Degree of polymerization, 126, 162, 227
Denatured alcohol, 17
Depolymerization, of methacrylate polymers, 156, 195
De-swelling, 54, 55, 97
Detection of cross-linking, 156, 164, 195-200 *passim*, 203-5, 213-17 *passim*
Deterioration. *See also* Cracks; Degradation; Discoloration; Exposure; Ultraviolet radiation
  by oxidation, 137, 151, 155, 159
  processes of, 6, 154-64 *passim*, 173, 211-12, 216-17
  stages of, 5, 154-64 *passim*, 202-8 *passim*, 211-12
Diacetone alcohol, 20, 26, 39, 89, 91, 92, 178, 180, 231
  discoloration of, 21, 137, 141
Dibutylphthalate, 130, 153
Dicarboxylic acids, 53, 71, 72, 75, 79, 80-85 *passim*, 87, 88, 108. *See also* Adipic acid; Azelaic acid; Pimelic acid; Suberic acid
  listed, 76
para-Dichlorobenzene insecticide, 11
Diethyl benzene, 37, 38, 91, 92, 99, 103, 110, 231
Diethylene glycol succinate (DEGS), 76
Diethyl ether, 18, 20, 26, 91, 231
Diffusion coefficients, 99-105 *passim*
  for solvent swelling and de-swelling, 103
Diffusion, of solvents, 33, 50, 55, 187. *See also* Diffusion coefficients.
  Fick's law of, 95, 100
2,4-Dihydroxybenzophenone, 199-200
Di-isopropyl benzene, 38, 91, 232
Diluent, solvent in cleaning pictures, 50
Dimethyl formamide, 91, 232
Dioctylphthalate, 130
1,4-Dioxane, 91, 232
Dipentene, 91, 232
Dipole forces, 32, 33, 40
Dipole moment, 28, 29, 32
Dipterocarpaceae, source of dammar, 120
Dirt
  exclusion of, 147, 152
  pick up, 146, 153

INDEX 239

Dirt (*continued*)
  removal before re-forming, 180
Discoloration
  influence of solvent on, 21, 124, 137, 139
  of poly(vinylidene chloride), 6, 113
  of turpentine, 13, 124, 155
  of varnish, 5, 6, 124, 139, 143-45, 148, 149, 152, 155, 173, 179, 181, 182
Disintegration, of varnishes, 154, 155
Dispersing agents, in emulsions, 218, 219, 222
Dispersion, 24, 29
Distinctness-of-image gloss, 139, 141
Divinylbenzene, 203, 205
DP (degree of polymerization), 126
n-Dodecane, used in solubility grade test, 33-36 *passim*, 132
Double bond, 10, 14, 126
Drying, 3-5, 137-39
  of polymer emulsions, 219-220
Drying oils, 17, 48, 51, 52, 54. *See also* Linseed oil films
Dull finish, 141
Du Pont polymers. *See* Polymers: Elvacite®
Durability. *See* Deterioration
  influenced by monomer, 131
Dust, 180
  pick up by varnish, 146, 153
Dyes, fading of, 158, 196

Ease of removal. *See* Removal
East Indies, source of dammar, 120
Elasticity, 123, 147, 152
Elongation-at-break, 127, 129, 147, 148. *See also* Brittleness
Elvacite®. *See* Polymers
Embrittlement, 155, 217. *See also* Brittleness
Emulsion polymerization. *See* Polymer emulsions
Esters, 18, 19, 20, 32, 40
  of glycerine, 18. *See also* Drying oils; Triglycerides
Ethanol, 15, 16, 17, 26, 39, 89, 178
  diffusion rate and coefficients of, 99, 102-4
  evaporation time of, 26, 110, 232

  leached components analyzed, 72
  leaching and swelling by, 59, 62, 64, 65, 67, 69, 106, 110, 111
  radioactive, 181-88 *passim*
  in removal of dammar varnish, 47
  in removal of oil films, 49-50
  solubility parameter of, 91-93
  toxicity and safety of, 223
Ethanolamine, 21
Ether. *See* Diethyl ether
Ethyl acetate, 18, 19, 22, 40, 91, 232
Ethyl alcohol. *See* Ethanol
Ethyl benzene, 89, 91, 92, 99, 103, 110, 232
Ethylene dichloride
  diffusion rate and coefficients of, 102, 103
  evaporation time of, 110, 232
  leaching and swelling by, 83, 107, 109-11, 174, 176
  in removal of dammar varnish, 49
  solubility parameter of, 91-92
  toxicity and safety of, 232
Ethylene glycol, 222
Ethylenic compounds, 14
Evaporation rate, 25, 39, 40, 103, 231-34
  influence on appearance, 139, 141
  influence on brittleness, 137
  of re-forming solvent, 181-88 *passim*
Evaporation retarder, 222
Evaporation time. *See also* Evaporation rate
  relative tables of, 26, 231-34
Experimental varnish 27H. *See* Polymers: poly(isoamyl methacrylate)
Exposure. *See also* Daylight; Fadeometer; Radiation; Ultra-violet radiation
  effects of. *See* Chain breaking; Cross-linking; Discoloration; Disintegration; Embrittlement; Removal; etc.

F-10. *See* Polymers: poly(n-butyl methacrylate)
Fadeometer. *See also* Cross-linking
  exposures in, 157, 158, 161, 195, 196, 202-6 *passim*, 211-17 *passim*
  fading of dyes in, 158, 196
Faraday-Tyndall effect, 24

Film. *See also* Linseed oil films
  formation in polymer emulsions, 219-21
  life of, 3-6 *passim*
  periods in deterioration of, 202-8 *passim*
  properties. *See* Absorption spectra; Brittleness; Color; Permeability; Polymers; Varnish; etc.
  repair of polymer, 141
  structure of, 188
Filters
  Corex D, 196
  National Bureau of Standards yellow, 148, 149
  window glass as, 149
  for ultra-violet light. *See* Ultra-violet absorbers; Varnish, as light filter
Filter paper, used in chromatography, 72, 197
Fire hazard, of solvents, 38, 231-34 *passim*
Flame ionization detector, 74
Flexibility, 4, 127. *See also* Brittleness
Flow, plastic, 119
Fluorescent lamplight, 149, 168
Fluorohydrocarbons, 36, 131
Fogg Art Museum
  early experiments with synthetic resins at, 125, 136, 228
  exposed samples at, 158, 159, 171, 172, 197, 200, 207, 209
Foil. *See* Aluminum foil
Formic acid, 17, 18
Formulation, of solvent, 138, 139, 141, 142
Fossil resins, 120

Gallery, exposure in. *See* Cross-linking, under gallery conditions
Gallery tone, of varnish, 124
Gas chromatography
  description of, 73-75
  in analysis of leached components, 83-86 *passim*
Gel
  dose, 162, 163, 205, 213, 215
  ease of removal as, 154, 159, 162, 189, 203-8 *passim*
  influence on appearance, 140
  shrinkage of, 154
  swelling of, 154, 163, 200

  syneresis of, 6, 154
  varnish as, 119
Gelva® poly(vinyl acetates), 227-28
Geneva nomenclature, 16, 18
Glass
  support for test films, 130, 225
  window, as filter, 149
Glass transition. *See* Second-order-transition temperature
Glassy state, 146, 165
Glazes, 141, 144
Gloss, 5, 138-41, 143, 152, 173, 223. *See also* Matte
Glycerine. *See* Glycerol
Glycerol, 18, 91, 232
Grain alcohol, 16. *See also* Ethanol
Ground, 150, 225
Gum turpentine. *See* Turpentine

Halogenated hydrocarbons, 21, 32, 36, 40, 131. *See also* Fluorohydrocarbons *and specific compounds, such as* Carbon tetrachloride
Halogens, 21
Hardness, 130, 131, 133, 139, 146, 188, 224
  effect of retained solvent on, 5, 6, 139
  of various resins, 122, 130, 132
Harrison, L. S., report on damaging effects of illumination, 149, 153, 196
Harvard. *See* Fogg Art Museum
Hazard of solvents. *See also* Toxicity
  in fire, 38, 231-34 *passim*
Heat, cross-linking by, 156, 195
n-Heptane, 89, 91, 92, 102, 104, 232
n-Hexane, 9, 22, 89
  diffusion rate and coefficients of, 102, 104
  leached components analyzed, 70, 79-81
  leaching and swelling by, 109
  solubility parameter of and swelling by, 91-94 *passim*
High molecular weight coatings, 8
  leveling of, 140, 141
  re-forming of, 141
Hoechst, poly(vinyl acetate) from, 227
"Hold out," of emulsion coatings, 223
Homologous compounds, 16, 17, 22, 40, 143

Howards of Ilford, 128
Hue, effect of varnish on, 143
Hydrocarbons. *See also* Aromatic hydrocarbons; Cycloparaffins; Paraffin hydrocarbons
  aromatic, definition of, 8, 10
  Kauri-Butanol number of, 36
  in petroleum, 11, 12, 36
Hydrogen, 7, 8
  tertiary, as active site, 157, 158, 166
Hydrogenated resins, 120. *See also* MS-2
Hydrogen bonding
  factor in solubility, 27, 28, 40
  tendency of various compounds, 30, 32, 33
Hydrogen peroxide, as oxidizing agent, 155
Hydroperoxides, 52
2-Hydroxy-4-methoxybenzophenone, as ultra-violet absorber, 149, 160, 199, 200

Illumination level, in long exposure to daylight, 154, 178, 196
Imperial Chemical Industries' Bedacryl 122X, 227
Incandescent lamplight, damage per footcandle of, 149
Indentation hardness, 131
Indonesia, source of dammar, 120
Induction period, 159, 207
  in cross-linking, 160, 202, 208. *See also* Gel, dose
Infra-red spectra,
  of resins, 123, 155, 183
  used in studies of leaching, 70–72
  used in studies of re-forming process, 181–85 *passim*, '187, 188
Inhibition, 158, 159, 160, 199, 200, 203
Insolubility, defined, 203
Insoluble matter. *See also* Removal
  effect on removal of, 196, 204–6
  increase in: laws governing, 162, 205, 213, 217, owing to oxidation, 212
  in methacrylate coatings from Fogg Art Museum, 158, 159
  period of increasing, upon exposure, 160–62, 202–7 *passim*, 212–17 *passim*
  tests for, 198, 203, 204, 208, 209, 212, 213
  toleration limit of, 154, 207

Intensity of light, effect on exposure rates, 164, 178, 196, 211
Internally plasticized polymer, 130, 153
Intrinsic viscosity, 126
Iodine value, 52
Isobutyl alcohol, 89, 90, 91, 92, 102, 104, 231
Isobutyl methacrylate. *See* Polymers: poly(isobutyl methacrylate)
Isomers, 10
Isooctane
  diffusion rate and coefficients of, 102–4
  evaporation rate of, 110, 233
  leaching and swelling by, 89, 110
  solubility parameter of, 91–92
Isopropanol, 15, 16, 89
  diffusion rate and coefficients of, 102–4
  evaporation time of, 110, 233
  leached components analyzed, 70, 72, 80
  leaching and swelling by, 59, 64, 65, 67, 69, 110
  solubility parameter of, 90–92
Isopropyl alcohol. *See* Isopropanol

Jelly. *See* Gel

Kauri-Butanol number, 12, 34, 36
Kauri gum, 34
K-B number. *See* Kauri-Butanol number
Kerosene, 12, 17, 91, 232
Ketones, structure of, 18, 20, 21, 32
Ketonic groups, formed upon oxidation, 121, 161

Lacquer, 3
Laurie, A. P., on solvents, 12
Law, mathematical, for decrease in solubility, 162, 205, 217
Leached components, quantitative analysis of, 80–88 *passim*
Leaching process, 66–68
  in dried linseed oil films, 54–60 *passim*
  effect of time of solvent contact on, 62, 63
  influence of solvent and film type on, 63, 64, 65, 66, 69
  and repeated swelling of films, 60–62
  soluble components: from, 68–70; studied by gas chromatography, 73–88; studied by infra-red spectroscopy, 70–72

Leaching test for cross-linking. See Chromatography
Leveling of varnish, 140, 141
Level of illumination. See Illumination level
Light
  action on poly(vinylidene choride), 133
  activation by, 156
  bleaching effect on varnishes, 124
  carbon-arc. See Fadeometer
  effect on polymers. See Chain breaking; Cross-linking; Daylight; Degradation; Discoloration; Exposure
  factors of damage per footcandle by, 149, 153. See also Harrison
  fading of dyes by. See Fadeometer
  in galleries. See Cross-linking, under gallery conditions; Illumination level; Intensity of light
  intensity increased by use of aluminum foil, 211
  oxidation induced by. See Oxidation
  transmission by varnish, 148. See also Color; Discoloration; Ultra-violet absorbers; Varnish, as light filter; Yellow
Linear polymers, 126, 130, 131, 155, 156, 207. See also Polymers, thermoplastic
Linoleic acid, 51
Linolenic acid, 17, 51
Linseed oil films
  diffusion solvents into, 95-105 passim
  drying of, 14, 50-54 passim, 159
  leached components analyzed by, 70-88 passim
  leaching and swelling of, 45, 54-70 passim, 89-90
  permeability of, 151
  solubility parameter and swelling of, 90-95 passim
  ultra-violet aging of, 64, 65, 66, 72, 83-87 passim, 156
Loss of solubility. See Cross-linking; Deterioration; Gel, dose; Insoluble matter; Oxidation
Low molecular weight polymers. See also AW-2 resin; MS-2; Viscosity grade
  as film formers, 128, 160
Lucanus, F., introduction of dammar by, 120

Lucas, A., introduction of nitrocellulose by, 125
Lucite® 44. See Polymers: poly(n-butyl methacrylate), Elvacite® 2044
Lucite® 45. See Polymers: poly(isobutyl methacrylate), Elvacite® 2045
Malaya, source of dammar, 120
Mandrel-bend test, 123, 124, 127, 128, 138. See also Brittleness
Manual on the Conservation of Paintings, 141
Mastic resin, 120, 121
  brittleness of, 123, 124, 127, 128
  effect of light on, 121, 124, 160
  hardness of, 122, 130
Mastic varnish, 120, 121, 123, 151, 171
  appearance of, 124, 141-45 passim, 150, 154, 155, 173
  brittleness of, 128, 137, 138, 148
  effects of age and re-forming of, 175, 176, 184-87 passim
  loss of radioactive ethanol from, 184-87 passim
  oxidation of, 121, 155, 160, 161, 217
  removal of, 47, 48, 121, 124
  solvent retained by, 137, 138
  transmission of light by, 148, 149, 150
  viscosity of, 121-23, 139-42 passim
Matte, surface, 139, 141
Maximum allowable concentrations. See Toxicity
Medium oil alkyd, solubility parameter of, 28
Metals, protection from corrosion, 150, 166
Methacrylate esters, 14
Methacrylate polymers. See Polymers
Methane, 8, 9, 32
Methanol, 15-17, 89, 174, 176
  diffusion rate and coefficients of, 102-4, 106, 107
  evaporation time of, 26, 110, 232
  leached components analyzed by, 79-80, 82-88 passim
  leaching and swelling by, 57-59, 62, 64-66, 69-70, 109-10
  solubility parameter of and swelling by, 90-98 passim
  toxicity and safety of, 16, 39, 40, 232

INDEX 243

Methyl acetate, 19, 91, 232
Methyl alcohol. *See* Methanol
Methyl cellosolve, 62, 69, 91, 103, 110, 232
Methyl cyclohexane, 153, 158, 159, 165, 209
Methylene chloride
   diffusion rate and coefficients of, 100, 103, 105
   evaporation time of, 110, 232
   leached components analyzed, 80, 82–85 *passim*, 88
   leaching and swelling by, 65, 68–70, 110
   solubility parameter of, 91
   toxicity and safety of, 232
Methyl esters of fatty acids, 75
Methyl ethyl ketone, 89, 91, 92, 93, 232
Methyl isobutyl ketone, 91, 232
Methyl methacrylate/ethyl acrylate copolymers. *See* Acryloid®; Rhoplex® AC-33
Methyl n-propyl ketone, 89, 91, 92, 232
Microcrystalline wax, 9
Migration, 130, 149
   in chromatography, 197
Mildew, 222
"Mild" solvent. *See* Solvent, power; Solubility, parameter; Removal
Mineral spirits, 12
Minimum film formation temperature, 221, 224
Mixtures of solvents
   formulations with, 138, 139
   in re-forming, 178
   in safety margin test, 121
   in solubility grade test, 34–35
   as standard of solvent strength, 153, 165
   swelling action of, 93–95
Modulus of elasticity, 123, 131
Mold, 222
Molecular weight, 8
   average, 126, 228
   effects of, 126, 140, 143, 160. *See also* Tensile strength
Molecular volume, 40, 89, 90
Molecules, 7, 8
   forces between, 27, 32
Monitoring of changes in coatings, 164

Monocarboxylic acids, listed, 76
Monomer, 119, 126, 131, 166
Morpholine, 181, 232
Mowilith poly(vinyl acetate), 227
MS-2, MS-2A, and MS-2B resins, 128, 129, 133, 154, 160

Naphtha, petroleum, 12, 36, 91, 174, 178, 197, 208, 233
Naphthenes, 12
National Accelerated Fading Unit, 196. *See also* Fadeometer
National Bureau of Standards
   research on cross-linking, 156
   yellow filter, 148, 149
Natural polymers, 119–21
Natural resins. *See also* Dammar resin; Mastic resin; Shellac; etc.
   available in spray cans, 125
   classification of, 120
   oxidation of, 121, 155, 161, 217
Nitrocellulose. *See* Cellulose nitrate
Nitrogen, compounds containing, 21, 30, 32, 120
Non-drying oils, 52
Non-swelling period in aging, 202, 206, 207
Non-volatile component, 5, 6, 25, 119–33 *passim*
Normal hydrocarbons, 10
Nylon, 28, 126

n-Octanol, 91, 233
Oil paints. *See* Paint; Paintings
Oil-resin varnishes, 120, 164, 177
Oils. *See* Linseed oil films
   drying, 17, 188
   non-drying, 52
Olefins, 14
Oleic acid, 17, 51
Oleoresinous films, 50. *See also* Oil-resin varnishes
Organic acids, 16. *See also* Dicarboxylic acids *and specific names*
Organic chemistry and compounds, 7, 14, 16, 17, 21
Outdoor exposure, 158, 196. *See also* Daylight

Oxidation
  of coatings, 6, 14, 155, 156, 189, 202, 212, 216, 217
  definition of, 155
  increases solubility parameter, 161, 163, 212, 217
  inhibition of, 159–60, 200
  loss of solubility owing to, 121, 161, 189, 212
  process of, 5, 150, 151, 159–60
  of solvent. *See* Discoloration; Turpentine;
Oxygen, 8, 30
  permeability to, 150, 151
Ozokerite wax, 9
Ozone, 155

Paint, 178, 189
  acrylic emulsion, 221
  chalking of, 154
  permeability of, 151
  stress-strain properties of, 147, 148. *See also* Linseed oil films
Painting Materials, 120
Paintings, analysis of specimens from, 66, 85, 86
Paint surface
  solvent attack on, 180
  varnish conformity to, 141
  wetting, by varnish, 142–44
Paint Thinners from Petroleum, 12
Palmitic acid, 17, 51
Paper, 151, 155, 223
  chromatography: in study of triglycerides, 72; test for cross-linking of, 158, 197–200 *passim*, 204, 209
  clips: analogy to monomer, 126
Paraffin hydrocarbons, 8, 9, 29, 34, 40
  wax of, 9, 151
  wetting by, 142
Parameter. *See* Solubility, parameter
Particle size of emulsions, 222–24
Peaks in gas chromatography, 76
Pencil hardness test, 131
Penetration
  of solvent. *See* Diffusion
  of varnish, 142, 143, 144, 181, 223
Periods
  in deterioration of polymers, 202–8 *passim*
  in life of solvent-type, 4–6

Permeability, 133, 150–52, 166, 222
Petroleum ether, 91, 233
Petroleum fractions. *See also* Aromatic hydrocarbons; Cycloparaffins; Naphtha; Paraffin hydrocarbons; Restrainer; White spirit
  solvent power of, 12, 22, 34, 36
  in 27H varnish, 208
Petroleum Thinner Index, 12, 36
Pettenkofer process, 49, 179
pH, of polymer emulsions, 219, 224
Phorone, cause of yellow color, 20–21, 155
Physical changes of film, 6, 154, 181, 184. *See also* Brittleness; Cracks; Disintegration; Gel; Shrinkage; Wrinkling
Physical properties of solvents and liquids, 230–34 *passim*
Picture varnish. *See* Varnish
Pigmented linseed oil films, 55–60 *passim*
Pigment-medium interaction, 60
Pimelic acid, 53
α-Pinene, 13–14
Pistacia Lentiscus, source of mastic, 120
Plastic and plastic flow, 119
Plasticization, internal, 130, 153
Plasticizer, 30, 119, 129, 130, 153
  in polymer emulsions, 218, 220, 221, 224
Plexiglas®, UF-1, ultra-violet filter, 149
Polarity of solvent, 27, 143, 180
Polar molecules. *See* Dipole moment
Polymer emulsions, 218–24 *passim*
  checklist of key properties of, 224
  drying of, 219–21
  minimum film-forming temperature of, 221
  mold prevention in, 222
  particle size of, 222–24
  pH of, 219, 224
Polymerization. *See* Addition polymer; Condensation polymers; Polymer emulsions
  degree of, defined, 126
  of linseed oil, 14, 52
Polymers (General)
  acrylic, 125, 126, 133, 158, 162, 195, 221
  condensation, 18, 126
  introduction of, in conservation, 125, 127, 129, 130, 164, 228

INDEX 245

Polymers (General) (*continued*)
methacrylate, 125, 142, 148, 149, 151, 153, 155, 156, 161, 173, 200, 211
natural, 119, 120, 121
polyester, 18, 126
thermoplastic. *See also specific types and subjects*
  classification of, 130, 131
  cross-linking and deterioration of, 154–64 *passim*, 195–217 *passim*
  cross-linking, resistant types, 157–58, 160, 196, 200
  in dammar, 121
  effects of exposure on. *See* Chain breaking; Cracks; Cross-linking; Degradation; Discoloration; Disintegration; Embrittlement; Exposure; Shrinkage; Weight loss
  internally plasticized, 130, 153
  linear chain concept of, 126, 130, 131, 155, 156, 188, 207
  molecular weight of. *See* Molecular weight
  oxidation of. *See* Oxidation
  solubility of, 27–33 *passim*. *See also* Formulation; Removal; Soluble fraction
  solubility grade of, 34, 35
  tensile strength of, 126, 162. *See also* Brittleness; Tensile strength
  viscosity grade of, 34, 132, 141, 227
thermosetting, 125, 126, 131, 156, 175, 188, 218
Polymers (Specific)
poly(n-amyl methacrylate), 35
polybutadiene, 205
poly(t-butyl acrylate), 195
poly(n-butyl-co-isobutyl methacrylate): Elvacite® 2046, 159, 197
poly(n-butyl methacrylate)
  Acryloid® F-10, 122, 129, 132
  Bedacryl 122X, 227
  brittleness of, 124, 126
  cross-linking of, 161, 166, 202–7 *passim*, 215: in daylight, 158, 159, 164, 197, 207, 209
  degradation weight-loss of, 161, 206
  dirt collection by, 146
  drying of turpentine-based varnish containing, 5, 155

Elvacite® 2044 (formerly Lucite® 44), 122, 123, 124, 127, 132, 136, 159, 161, 202, 227
  effect of exposure to fluorescent lamplight, 164
  hardness of, 122, 130
  infra-red spectroscopy after re-forming of, 182–85 *passim*
  oldest samples of (Fogg Art Museum), 125, 159, 172, 197, 200, 207, 209
  permeability of, 151
  removal of, 157–59, 162–64, 200, 204–6, 209
  solubility of before and after re-forming, 177, 178, 181
  solubility grade of, 5, 34, 35
  solubility parameter of, 28
  viscosity grade of, 122, 123, 124, 127, 129, 130, 132, 227
polycyclohexanone. *See* AW-2 resin; MS-2
poly(cyclohexyl methacrylate), 35
poly(2-ethylbutyl methacrylate), 157
polyethylene, cross-linking of, 156
poly(ethylene glycol), 153
poly(ethyl methacrylate), 34–35, 125, 129, 130
poly(isoamyl methacrylate)
  brittleness of, 124, 128
  cross-linking of, 156–58, 195–200 *passim*, 202–8 *passim*:
    in daylight, 158, 159, 196, 197, 200; inhibited, 160, 199, 200
  degradation and depolymerization of, 195, 206
  experimental varnish 27H, 4, 128, 147, 148, 149, 159, 190, 191, 202, 203, 204, 208
  hardness of, 130, 132
  permeability of, 151
  removal of, 157–59, 162, 195, 197, 200, 203–6 *passim*
  solubility grade of, 34, 35
  viscosity grade of, 124, 128, 130, 208
poly(isobutyl methacrylate)
  Acryloid® B-67, 34, 122, 127, 129, 132
  brittleness of, 124
  cross-linking of, 213, 214: in daylight, 158, 159, 164, 196, 197, 200

poly(isobutyl methaerylate) (*continued*)
  Elvacite® 2045 (formerly Lucite® 45), 122, 124, 127, 132, 159
  hardness of. 130, 132
  long-term use in conservation, 125, 127, 197
  removal of, 157-59, 200
  solubility grade of, 34, 35
  solubility parameter of, 28
  viscosity grade of, 122, 124, 130, 132
poly(2-methylbutyl methacrylate) and poly(3-methylbutyl methacrylate), 157
poly(p-methylcyclohexyl-co-isoamyl methacrylate), 161, 213
poly(methyl methacrylate), 29, 30, 34, 35, 125, 151, 195, 205
poly(methyl methacrylate-co-ethyl acrylate), 132, 158, 161, 200, 221, 222
  Acryloid® B-72, 34, 37, 124, 125, 127, 129, 132, 161, 200, 212
  Acryloid® B-82, 161, 212
  Rhoplex® AC-33, 161, 212, 221
  Rhoplex® AC-55, 221
poly(neopentyl methacrylate), 35
poly(n-propyl methacrylate), 34, 35, 130, 157, 158, 161, 175, 178
polystyrene, 133, 151, 172:
  brittleness of, 124, 127, 128;
  cross-linking of, 156, 203, 205
poly(vinyl acetal): Alvar®, re-forming of, 174, 177
poly(vinyl acetate), 14, 22, 125-27, 133, 137, 153
  deterioration resistance of, 120, 133, 158, 173, 196
  formulation of, 141, 146
  hardness of, 122, 130, 132
  light transmission of, 148
  loss of radioactive ethanol from, 185-88
  molecular weights of, 126, 228
  permeability of, 151
  proprietary types, 226-28. *See also this entry*: Union Carbide products *and* Vinylite A
  re-forming of, 175, 177, 184, 185, 187
  solubility of, 5, 22, 175, 177
  solubility grade of, 34, 132

Union Carbide products. *See also this entry*: Vinylite A; VYHH
  AYAA, 122-24, 128, 132, 227
  AYAB, 122-24, 127, 132, 142, 146, 227
  AYAC, 132, 146, 227
  AYAF, 4, 122, 123, 127, 132, 138, 141, 142, 227
  AYAT, 122, 123, 127, 142, 227
  used in retouching, 21, 146
  Vinylite A, 172, 173, 177, 184, 185, 187, 228
  viscosity grades of, 132, 227
poly(vinyl alcohol), 142, 153
poly(vinyl chloride), 31
poly(vinyl chloride-acetate): Union Carbide product VYHH, 4, 129, 151, 172, 173, 177
poly(vinylidene chloride): Saran®, 6, 133, 151
Pressurized containers for spraying, 125
n-Propanol, 16, 17, 89, 90, 91, 92, 102, 103, 104, 110, 178, 180, 233
  diffusion rate and coefficients of, 102-4
  evaporation time of, 110, 233
  leaching and swelling by, 110
  solubility parameter of, 90-92
Properties. *See* Absorption spectra; Brittleness; Color; Permeability; etc.
Proprietary polymers. *See specific trade designations, as* Acryloid®; AW-2 resin; Terpolymer A
  formulations of, 125, 209
n-Propyl acetate, 91, 233
n-Propyl alcohol. *See* n-Propanol
Propylene glycol, 222
Protective coating. *See* Polymers; Varnish, protection by
Protein, 32, 119, 126
Pyridine, 21, 32, 91, 233

Radiation. *See* Daylight; Fadeometer; Infra-red spectra; Ultra-violet radiation
  protection from, 148-50, 152, 153
Radioactive ethanol, 184-88 *passim*
Radicals, 15, 158, 160
Random distribution of triglycerides, 78

INDEX 247

Ratio of chain breaking to cross-linking, 162, 163, 211–17 *passim*
Reagent grade solvent, 25
Re-forming
  effect on ease of removal, 171, 175, 176, 181
  history of technique, 49, 179
  infra-red spectroscopy of process, 181–85 *passim*
  mechanism explained, 188–89
  mechanism investigated with radioactive ethanol, 185–88 *passim*
  solvent for, 178
  technique of, 141, 178–79, 180–81
Refractive index, 12, 127, 133, 142, 143, 144
Relative evaporating time. *See* Evaporation time
Relative retention ratios, 74
Removal, of varnish. *See also* Cross-linking; Non-swelling period in aging; Oxidation
  ease of, 152–54, 202–8 *passim*. *See also* Re-forming
  facilitated by brittleness, 164, 206
  frequency of, 147–48
  influenced by swelling, 162, 163, 187, 205, 206
  of long-term exposed polymers, 158, 159, 200, 209
  of natural resins, 121, 124, 165, 175, 176, 190, 209, 210
  in relation to gel dose, 163
  resistance to, defined, 157, 196, 208
  as sol or gel, 162, 189, 202, 204
  solvent "strength" in, 121, 147, 153, 163, 200, 206
  time of. *See* Re-forming; Swabs, tests using
Repair of varnish surface, 141
β-Resene, 121. *See also* Dammar wax
Resin, definition and specification of, 119, 153
Resin-oil varnishes. *See* Oil-resin varnishes
Resins. *See specific names, as* AW-2 resin; Dammar resin; Mastic resin; MS-2; Shellac
Restrainer, 12, 48, 50

Retained solvent, 3–5, 137–39, 155, 181, 182, 185. *See also* Ethanol, radioactive; Re-forming
Retouching, 21, 129, 146, 154
R_F, 198–200. *See also* Chromatography, paper: test for cross-linking
Rhodiatoce, Societa, 227
Rhodopas, poly(vinyl acetate), 227
Rhone-Poulenc, Société des Usines Chimiques, 227
Rhoplex® AC-33, 161, 212, 221
Rhoplex® AC-55, 221
Rocker hardness, 131
Rohm and Haas, 122, 132, 221. *See also* Acryloid®; Rhoplex® AC-33; Rhoplex® AC-55; Plexiglas®
Rome Conference
  of 1930, 125, 152
  of 1961, 212
Rosin. *See* Colophony
Rough surface, 138, 140, 141, 142, 144

Safety, 38–40. *See also* Fire hazard; Toxicity
Safety margin test, 50, 121
Sandarac, 121
Saponification technique, 75
Saran®. *See* Polymers: poly(vinylidene chloride)
Saturated fatty acids in oils, 51, 70–73 *passim*
Saturated hydrocarbons. *See* Paraffin hydrocarbons
Scission reactions, 53. *See also* Chain breaking; Degradation; Weight loss.
Scratch hardness, 131
Second-order-transition temperature, 119, 146, 147, 153, 221, 222, 224
Semi-drying oils, 52
Shawinigan, 227
Sheen, 139
Shellac, 121
Shell sol 715 (72), 91, 95, 233
Shrinkage, 58, 59, 60, 154, 161
Silica gel, in test for aromatics, 12
Silicones, 120, 131
Skylight. *See* Daylight; Illumination level

Slow-evaporating solvents, use of, 138, 141, 142
Soaps, 17, 146
Sodium methoxide technique, 75
Softening temperature, 126
Sol, 162, 202, 204
Solubility. *See* Cross-linking; Insoluble matter; Soluble fraction
   of aged solvent-type coatings. *See* Deterioration; Insoluble matter; Leaching process; Re-forming; Removal; Soluble fraction
   grade, 33–35 *passim*, 40, 131–32, 133
   of linseed oil films, 59
   of natural resins, 120
   parameter
      of aged solvent-type coatings, 203, 210, 212, 216
      of dried linseed oil films, 92. *See also* Swelling
      as specification, 22, 27–34 *passim*, 40, 90–95 *passim*, 131
      of various solvents, 91
      of polymers (unexposed), 28, 30–33 *passim*
      of polymers vs. removability, 202–8 *passim*
   rate of, defined, 189
Soluble fraction, 204, 212, 214. *See also* Insoluble matter
Solute, 25
Solution, 24–27 *passim*, 32, 33, 189
Solvent. *See also* Removal *and specific names*
   classification of, 25, 32, 40, 139
   cohesive energy density of. *See* Solubility, parameter
   definition of, 24, 25
   effect on film properties. *See* Discoloration; Formulation; Gloss; Retained solvent; Spray application
   evaporating rate or time of, 25, 26, 39
   formulation of, 138, 139, 141, 142
   grades of purity of, 25
   hydrogen bonding, tendency of, 27–33 *passim*
   oxidation of. *See* Turpentine, discoloration
   polarity and dipole moment of, 27–29, 32
   power, 22, 142, 153
      standards of, 121, 143, 153, 165. *See also* Solubility, grade *and* parameter
   rate of attack on films, 26, 33, 171–78 *passim*, 180, 181, 189
   re-forming mixture, 178
   retained, 3–5, 137–39, 155
   risk of action by, 33, 141, 148
   safety of, 25, 38, 39. *See also* Fire hazard; Toxicity
   safety margin test, 121
   swelling by. *See* Swelling
   toxicity, 38, 231–34 *passim*
   vapors, 25. *See also* Fire hazard; Toxicity
Solvent action, 12, 25, 26, 36, 51, 141, 148
Solvent-type varnish. *See* Varnish
Specifications
   of petroleum fractions, 12
   for picture varnish, 125, 152, 153
   of properties, thermoplastic polymers, 133
   of toleration limit of insoluble matter, 154, 207
   of yellow color (standards), 145
Specular gloss, 139
Spirit, British white. *See* White spirit
Spirits, mineral, 3, 12
Spirit varnish, 3, 171, 172. *See also* Varnish, solvent-type
Spray application
   of re-forming solvents, 179–89 *passim*
   of varnish, 125, 138, 141, 144
Stable polymers, 120, 133
Stages of deterioration. *See* Deterioration
Standards. *See* Solvent, power; Specifications; Stable polymers; Test
Stand oil, in dammar varnish, 123, 129
Stand oil films
   diffusion of solvents into, 96–98; 101–4 *passim*
   influence of molecular volume on swelling of, 89–90
   influence of solubility parameter on swelling of, 92–95 *passim*
   leaching and swelling of, 56–59 *passim*, 61–69 *passim*
Stationary phase, 74
Stearic acid, 17, 51
Stearyl alcohol, 15, 16

Storage of waste solvents, 38
"Strength" of solvent. *See* Solubility, parameter; Solvent, power; standards of
Stress, 5, 6, 119, 147, 154
Stringing of varnish, 138
Structure of compounds, 7-21 *passim. See also specific type*
Styrene, 14, 126
Suberic acid, 53
Substituted benzenes. *See* Alkyl benzenes
Substituted benzophenones, 199
Sulfur dioxide, 155
Sulfuric acid, test of aromatic content with, 12
Sunlight, 124. *See also* Daylight; Ultraviolet radiation
Surface of varnish, 5, 138, 139, 141, 143, 151, 152, 161, 164, 180. *See also* Bloom; Dust; Gloss; Spray application; Wetting; Wrinkling
Suspension, 24
Swabs, cotton, 174, 176, 177, 189
   storage of, 38
   tests using, 157, 163, 164, 196, 197, 200, 204-10 *passim*
Sward hardness, 122, 130-32
Swelling
   of cross-linked polymers, 126, 154, 156, 163, 200, 202-8 *passim*
   as a factor in re-forming, 187-89
   influence on leaching process, 69
   rate of linseed oil films, 96-98, 179
   stage in deterioration of cross-linked polymers, 163, 205, 207
Swelling action of linseed oil films
   description of, 54-60 *passim*
   equilibrium degree of, 60
   measurement of, 55
   modes of, 106-10 *passim*
   molecular volume of solvent and, 89-90
   solubility parameter of solvent and, 27, 33, 90-95 *passim*, 163
   virgin and leached films, 60-62
Syneresis, 6, 154
Synthetic, 7

Technical grade solvent, 25
Temperature
   effect of, on cross-linking, 162, 196, 215
   effect of, on diffusion coefficients, 105
   minimum film-forming, 221, 224
   second-order-transition, 119, 146, 147, 153, 221, 222
   thermal-distortion, 4
Tensile strength, 4, 126, 127, 219, 223, 224
   loss of, 162. *See also* Embrittlement
Terpenes. *See* Triterpenes; Turpentine
Terpolymer A, 213-16 *passim*
Tertiary hydrogen atoms, 157, 158, 166
Test
   accelerated aging, 162, 163, 208. *See also* Daylight; Exposure; Fadeometer
   for aromatic content, 12, 34
   using cotton swabs. *See* Swabs
   for cross-linking, 164, 196-98, 204, 209, 213-17 *passim*
   for ease of removal, 176
   for insoluble matter, 203, 208
   Kauri-Butanol, 34, 36
   mandrel-bend for brittleness, 127
   to monitor changes in coatings, 164
   pencil hardness, 131
   for pH of emulsions, 219
   solubility grade, 34
   of solvent "strength," 121
   viscosity grade, 120, 127
Tetrachloroethylene, 91, 233
Tetrahydrofuran, 91, 103, 110, 233
Textiles, 155, 223
Thermal-distortion temperature, 4
Thermally-processed natural resins, 120
Thermoplastic polymers. *See* Polymers, thermoplastic *and specific names*
   process of deterioration of, 156, 216. *See also* Cross-linking; Deterioration; Exposure; Gel, dose; Oxidation; etc.
   properties, 125-33 *passim. See also* Brittleness; Hardness; Viscosity grade
Thermosetting polymers, 125, 126, 131, 156, 175, 188, 218
Thickening agent, 218, 222, 223, 224
Time. *See* Diffusion; Evaporation time; Exposure; Removal
Tinted varnish. *See* Toned varnish
Tolerance of solvent. *See* Toxicity
Toleration limit of insoluble matter, 153, 154, 207
Toluene, 10, 11, 12, 26, 36, 38-40, 64, 67, 91, 233

Toluene (*continued*)
  in removal of cross-linked polymers, 157, 164, 195, 203-8 *passim*
  used in solubility grade test, 34, 35
  as solvent, 22, 33-36 *passim*, 121, 132, 137, 138, 143, 212, 221
  swelling by, 89, 92, 103, 182
  used in viscosity grade test, 122-24, 126-33 *passim*
Toluol, 13, 14. *See also* Toluene
Toned varnish, 124, 144. *See also* Color; Discoloration
Toxicity, of solvents, 38-40, 231-34 *passim*
Transmission
  of light. *See* Absorption spectra; Color; Ultra-violet absorbers; Varnish, as light filter
  of water vapor. *See* Permeability
Transparency, 152, 153
1,1,1-Trichloroethane, 21, 36
1,1,2-Trichloroethane, 91, 233
Trichloroethylene, 89, 91, 92, 110, 233
Triglycerides, 51, 54, 70-73 *passim*, 78
Triterpenes, 121
Turpentine, 13, 29, 36, 47, 49, 50, 89, 91, 92, 137, 233
  discoloration of, 13, 124, 155
  retained, 4, 5, 137, 138
27H. *See* Polymers: poly(isoamyl methacrylate)
Two-phase system, 6, 154

Ultra-violet absorbers, 149, 150, 199, 200.
Ultra-violet radiation
  aging of linseed oil films by, 64-66, 72, 83-87 *passim*
  effect on leached components, 83, 85, 86
  effect on natural resins, 124, 182
  effect on polymers, 164, 195, 198, 202, 207, 211, 216. *See also* Degradation; Discoloration; Exposure; Fadeometer
Unpigmented linseed oil films, swelling of, 55-60 *passim*
Unsaturated hydrocarbons and fatty acids, 14, 52
U.V. *See* Ultra-violet radiation
Uvinul®, ultra-violet absorber, 159

Valence, 8, 10, 21, 188
Value, 143. *See also* Color
Vapor
  barriers. *See* Permeability
  exposure to solvent, 180. *See also* Pettenkofer process; Toxicity
  pressure, 26, 38
Varnish
  air bubbles and poor wetting by, 143
  allowing color to be seen, 142
  appearance of, 5, 119, 138-44 *passim*, 152, 153, 173, 179. *See also* Bloom; Color; Gloss; Matte; Refractive index; Surface of varnish; Toned varnish; Wrinkling
  bleaching effect of light on, 124
  blooming and blanching of, 5, 124, 155
  color and discoloration of, 5, 6, 21, 124, 139, 142-45 *passim*, 148, 149, 152, 153, 155, 173, 182
  cracks in, 147, 152, 154, 161. *See also* Brittleness
  deterioration of, 6, 154. *See also* Cracks; Deterioration; Exposure; Oxidation; Shrinkage
  drying of, 3, 4, 137, 138
  elasticity of, 123, 147, 152. *See also* Hardness; Modulus of elasticity
  elongation-at-break of, 127, 148. *See also* Brittleness
  exposure of. *See* Daylight; Exposure; Fadeometer
  infra-red spectra, 123, 155, 183
  as light filter, 47, 143, 148-50, 152, 153, 181. *See also* Color; Discoloration
  natural resin. *See specific resins*
  oil-resin, 3
  optical properties of. *See* Color; Gloss; Matte; Refractive index
  oxidation of. *See* Oxidation
  penetration by, 142, 223
  permeability of, 150-52, 166
  proprietary, based on methacrylate polymers, 125, 209
  protection by, 146-52 *passim*, 166
  second-order-transition temperature of, 119, 146, 147, 153
  solubility of. *See* Cross-linking; Insoluble matter; Loss of solubility; Re-

INDEX   251

Varnish (*continued*)
   forming; Removal; Solubility, parameter; Soluble fraction
   solvent-type, 1–6 *passim*, 171
   specifications for, 152–54
   spirit. *See this entry*, solvent-type
   spray application of, 125, 138, 141
   viscosity, 122, 123, 138–44 *passim*
Ventilation, in use of solvents, 38
Vinalak: poly(vinyl acetate), 227
Vinavil: poly(vinyl acetate), 227
Vinyl acetate (monomer), 14, 28, 120, 126, 148. *See also* Polymers: poly(vinyl acetate)
Vinylite A. *See* Polymers: poly(vinyl acetate)
Vinyl Products, Ltd., 227
Viscometer, 55
Viscosity
   influence on appearance, 138–44 *passim*
   influence of solvent formulation on, 5, 122, 123, 138, 139
   of polymer emulsions, 219, 223
   in relation to diffusion coefficients, 104
   in solution, influenced by molecular weight, 126, 219
Viscosity grade, 122–33 *passim*
   definition of, 126, 127
   of low molecular weight resins, 124, 128, 138. *See also* AW-2 resin; MS-2
V. M. & P. Naphtha, 91, 176, 233
Volatile component, 3. *See also* Plasticizer; Solvent
Volatile plasticizers, 130, 153
Volatility, 8, 9, 25
VYHH. *See* Polymers: poly(vinyl chloride-acetate)

Wacker-Chemie GmbH, 227
Water, 8, 14, 15, 16, 18, 20, 21, 139, 142, 146, 147, 151
   hydrogen bonding in, 30, 32
   lost in condensation polymer, 126
   in polymer emulsions, 218–24 *passim*
   vapor, permeability to, 150, 222
Wax, 180
   beeswax, 41
   ceresin, 9
   dammar, 120, 121
   permeability of, 142, 151
   solubility of, 41
Wax-resin adhesive, 150
Weaver Report, 50
Weight loss, on exposure, 159, 161, 206, 208, 212, 216
Wetting, of paint surface, 142, 143, 144
White spirit (British Standard 245-1956), 12, 89, 91, 92, 94, 233
Window glass. *See* Glass
Wire screens, determination of insoluble matter in, 203, 208
Wood, 150, 223
Wrinkling, of varnish, 143, 154

Xylene, 10, 11, 12, 13, 22, 39, 91, 92, 103, 143, 153, 154, 197, 221
   evaporating time of, 26, 233
   Kauri-Butanol number of, 36
   as standard of solvent "strength," 121, 143, 153, 165
   toxicity and safety of, 39, 40

Yellow, color of varnish, 21, 139, 142–45 *passim*, 149, 153, 173, 182. *See also* Color; Discoloration
Yield value, 119

Zenith skylight, 149
Zinc white, action with dammar, 154

BIBLIOGRAPHY

The following is a selected bibliography of articles relating to the application of solvents, thermoplastic resins, and solvent-type varnishes that either had not been noted in the 1971 edition, or that have appeared since that date. The list was compiled from the Art and Archeology Technical Abstracts (AATA), particularly from the annotated bibliography by E. DeWitte and M. Goessens-Landrie on "The Use of Synthetic Polymers in Conservation," which appeared as a supplement in AATA, vol. 13, nos. 1 and 2, 1976. The source and abstract number of each reference is gratefully acknowledged as either coming from the AATA DeWitte, Goessens-Landrie bibliography (DW-GL), or from the AATA issues in the subsequent period, 1976 to 1983.

Some of the earliest references to the use of vinylacetate and methacrylate varnishes are cited, as well as a number of later reports describing conservation treatments in which these and other relevant thermoplastic resins have been used as surface coatings or retouching vehicles. Publications by conservation scientists especially concerned with the chemistry and physics of solvents and coatings are also noted. The bibliography has been added to the 1985 edition for the sake of completeness; inclusion in this listing is not to be taken as an endorsement of particular processes and materials described or of scientific conclusions reached.

1935 W. J. Clarke and H. E. Ives, "The Use of Polymerized Vinyl Acetate as an Artist's Medium," *Technical Studies in the Field of the Fine Arts* 4 (1935), 36-41 (AATA, DW-GL 9).

1937 G. L. Stout and H. F. Cross, "Properties of Surface Films," *Technical Studies in the Field of the Fine Arts* 5 (1937), 241-248 (AATA, DW-GL 16).

1940 S. Keck, "The Transfer of a Small Icon to a Support of Vinyl Resin," *Technical Studies in the Field of the Fine Arts* 9 (1940), 11-19 (AATA, DW-GL 22).

1950 T. R. Gairola, "Use of Methyl Methacrylate for Antiquities," *J. Indian Museums* 6 (1950), 40-41 (AATA, DW-GL 54).

1951 S. Keck, "A Case of Artistic Face-Lifting," *Brooklyn Museum Bulletin* 12 (1951), 16-21 (AATA, DW-GL 61).

S. Takakage, "Some Problems on the Preservation of Wall Paintings Using Synthetic Resins," *Sci. Pap. Japn. Antiq.* (1951), 29-31. In Japanese (AATA, DW-GL 63).

1956 P. Kostrov, "The Restoration of Two Fayum Portraits," *Soobshchenija Gosudartviennogo Ermitgea* (1956), 58-61. In Russian (AATA, DW-GL 137).

1957 R. D. Buck, "Oberlin's Ribera: A Case History," *Bull. Allen Mem. Mus.* 14 (1957), 69-72 (AATA, DW-GL 151).

T. R. Gairola, "Preservation of a Textile and a Miniature Painting," *Ancient India* 13 (1957), 143 (AATA, DW-GL 158).

M. Hamsik, "Some Questions of the Restoration and Scientific Examination of Paintings," *Zprävy Pamätkové Péce* 17 (1957), 107-111. In Czech (AATA, DW-GL 159).

1961 "Krylon Crystal Clear No. 1301," *Rohm and Haas Reporter* 19 (1961), 12 (AATA, DW-GL 259).

P. I. Kostrov and E. G. Sheinina, "Restoration of Monumental Painting on Loess Plaster Using Synthetic Resins," *Studies in Conservation* 6 (1961), 90-106 (AATA, DW GL 275).

1962 C. Steen and R.J. Gettens, "Tumacacori Interior Decorations," *Arizoniana* 3 (1962), 7-33 (AATA, DW-GL 337).

R.E. Straub, "Retouching with Synthetic Resin Paint," *Museum J.* 62 (1962), 113-119 (AATA, DW-GL 338).

1963 E.H. Jones, "The Effect of Aging and Reforming on the Ease of Solubility of Certain Resins," in *Recent Advances in Conservation* (London, 1963), 79-83 (AATA, DW-GL 368).

H. Lank, "MS2-A," *IIC News II* (1963), 8 (AATA, DW-GL 373).

R.E. Straub, "A Note on AW-2 and MS2-A," *Museum J.* 63 (1963), 123-124 (AATA, DW-GL 390).

G. Thomson, "A Note on MS2-A," *IIC News II* (1963), 13 (AATA, DW-GL 393).

1964 S. Keck, "The Past Preserved," *Arts in Virginia* 4 (1964), 27-30 (AATA, DW-GL 423).

1965 S. Keck, "The New York Conservation of a Peripatetic Portrait by John Singleton Copley," *Connoisseur* 159 (1965), 138-143 (AATA, DW-GL 478).

M. Ruggles, "Conservation of a Painting by the Transfer Method," *National Gallery of Canada Bulletin* 3 (1965), 26-31 (AATA, DW-GL 497).

K. Silberfeld, "The Lost Painting of Thomas Cole," *Museum News* 43 (1965), 31-38 (AATA, DW-GL 502).

1968 UNESCO, "Synthetic Materials Used in the Conservation of Cultural Properties," in *The Conservation of Cultural Properties with Special Reference to Tropical Conditions* (Paris, 1968), 303-327 (AATA, 15-195).

1970 B. Mühlethaler, "The Ability of Synthetics to Withstand Aging," *Arbeitsblätter für Restauratoren*, no. 2, group 16 (1970), 22-24. In German (AATA, DW-GL 746).

1971 E.C. Hulmer, "Notes on the Formulation and Application of Acrylic Coatings," *IIC-AG Bulletin* 11 (1971), 132-139 (AATA, DW-GL 786).

E. C. Hulmer, "Notes on the Formulation and Application of Adhesives and Supports," *IIC-AG Bulletin* 12 (1971), 46-54 (AATA, DW-GL 787).

1972 R. L. Feller, "Problems in the Investigation of Picture Varnishes," *Conservation of Paintings and the Graphic Arts*, IIC, Lisbon Congress (1972), 201-209 (AATA, DW-GL 825).

R. L. Feller and C. W. Bailie, "Solubility of Aged Coatings Based on Dammar, Mastic and Resin AW2," *IIC-AG Bulletin* 12 (1972), 72-81 (AATA, DW-GL 826).

E. C. Hulmer, "Notes on the Formulation and Application of Acrylic Coatings," *Conservation of Paintings and the Graphic Arts*, IIC, Lisbon Congress (1972), 211-213 (AATA, DW-GL 832).

H. Lank, "Picture Varnishes Formulated with Resin MS2A," *Conservation of Paintings and the Graphic Arts*, IIC, Lisbon Congress (1972), 215-216 (AATA, DW-GL 839).

1973 E. DeWitte, "The Protection of Silverware with Varnishes," *Bulletin Inst. Roy. Patr. Art.* 14 (1973), 140-151 (AATA, DW-GL 869).

G. Kaltenbrunner, "Acrylic Resin Colors in the Graphic Arts and Book Restoration," *Mitteilungen der Internationalen Arbeitsgemeinschaft der Archiv-, Bibliotheks- und Graphikrestauratoren* 3 (1973), 539. In German (AATA, 12-183).

1974 "Preserving a Priceless Art Treasure," *Rohm and Haas Reporter* 32 (1974), 1-10 (AATA, DW-GL 913).

J. Bernstein, "The Treatment of an Extensively Damaged Oil Painting on Canvas," *IIC-AG Bulletin* 14 (1974), 93-104. In German (AATA, DW-GL 918).

M. Dauchot-Dehon, "The Effects of Solvents on the Paint Layers. 1. Alcohols and Acetone," *Bulletin Inst. Roy. Patr. Art.* 14 (1973-1974), 89-104. In French (AATA, 12-979).

I.-L. Eckmann, "A Method of Treating Tented Lifting of Oil Paint on Canvas," *IIC-AG Bulletin* 14 (1974), 44-52 (AATA, DW-GL 929).

B. B. Lal, "Preservation of Mural Paintings," *Archaeological Survey of India, New Delhi. Ancient India,* no. 22-1966 (1974), 83-101 (AATA, 13-372).

1975 P. H. Brauer, "Modern Synthetic Varnish Resins for Use by Painters and Conservators," *Maltechnik/Restauro* 81 (1975), 96-98. In German (AATA, 12-974).

M. Curran, "Scattering of Light Over a Black Background by Matt Varnishes Based on Paraloid B-72," *ICOM Comm. for Cons.* (Venice, 1975), paper 75/22/3, 1-5 (AATA, 13-354).

E. DeWitte, "The Influence of Light on the Gloss of Matt Varnishes," *ICOM Comm. for Cons.* (Venice, 1975), paper 75/22/6, 1-9 (AATA, 13-356).

R. L. Feller, "Speeding Up Photochemical Deterioration," *Bulletin Inst. Roy. Patr. Art.* 15 (1975), 135-150 (AATA, 13-177).

R. L. Feller, "Studies on the Photochemical Stability of Thermoplastic Resins," *ICOM Comm. for Cons.* (Venice, 1975), paper 75/22/4, 1-11 (AATA, 13-641).

R. L. Feller and M. Curran, "Changes in Solubility and Removability of Varnish Resins with Age," *Bull. AIC* 15 (1975), 17-26 (AATA, 12-985).

N. G. Gerassimova, E. P. Melnikova, M. P. Rova, and E. G. Sheinina, "New Possibilities of (Poly) (Butyl Methacrylate) as a Consolidating Agent for Glue Painting on Loess Plaster," *ICOM Comm. for Cons.* (Venice, 1975), paper 75/1/4, 1-10. In English (AATA, 13-360).

N. Gerassimova, E. Melnikova, E. Sheinina, and M. Vinokurova, "New Ways of Using Poly Butyl Methacrylate as a Consolidating Agent," *Reports of the State Hermitage Museum* 40 (1975), 90-91. In Russian (AATA, 13-1017).

G. E. Mâle and D. Pigmerol, "Historical Study of Painting Varnishes from French Texts of 1620-1803," *ICOM Comm. for Cons.* (Venice, 1975), paper 75/22/1, 1-11. In French (AATA, 13-1305).

I. V. Nazarova, "Retouching Varnish for Paintings Based on Synthetic Resins," *ICOM Comm. for Cons.* (Venice, 1975), paper 75/22/8, 1-8. In French (AATA, 13-1043).

D. J. Newman and C. J. Nunn, "Solvent Retention in Organic Coatings," *Progress in Organic Coatings* 3 (1975), 221-243 (AATA, 13-1244).

K. Nitzi and M. C. Zech, "Infrared Spectroscopic Determination of Vinyl Acetate in Ethylene/Vinyl Acetate Copolymers," *Adhaesion* 19 (1975), 305-306, 308-309. In German (AATA, 13-1246).

1976 R. L. Feller, "Problems in the Investigations of Picture Varnishes," *Conserv. Restor. Pict. Art* (1976), 137-144 (AATA, 14-1011).

R. L. Feller, "The Relative Solvent Power Needed to Remove Various Aged Solvent-Type Coatings," *Conserv. Restor. Pict. Art* (1976), 158-161 (AATA, 14-1012).

N. I. Gaynes, "Acrylic Resins Are the Technology," *Metal Finishing* 74 (1976), 47-54 (AATA, 14-603).

E. C. Hulmer, "Notes on the Formulation and Application of Acrylic Coatings," *Conserv. Restor. Pict. Art* (1976), 145-147 (AATA, 14-1022).

R. H. Lafontaine, "Comparison of the Efficiency of Ultraviolet Absorbing Compounds," *AIC Bulletin* 16 (Winter 1975-1976), 74-94 (AATA, 13-107).

J. S. Mills, "The Identification of Paint Media—An Introduction," *Conserv. Restor. Pict. Art* (1976), 69-71 (AATA, 14-1041).

N. Stolow, "Solvent Action," *Conserv. Restor. Pict. Art* (1976), 153-157 (AATA, 14-1050).

1977   J.S. Mills and R. White, "Natural Resins of Art and Archaeology; Their Sources, Chemistry and Identification," *Studies in Conservation* 22 (1977), 12-31 (AATA, 14-1314).

L. Pomerantz, D. Goist, and R.L. Feller, "Conservators Advise Artists," *Art Journal* 37 (Fall 1977), 34-39 (AATA, 15-585).

1978   R.L. Feller, "Standards in the Evaluation of Thermoplastic Resins," *ICOM Comm. for Cons.* (Zagreb, 1978), paper 78/16/4, 1-11 (AATA, 16-130).

R.H. Lafontaine, "The Effect of Inhibitors on the Removability of Aged Ketone Resin N (BASF) Film," *J. of the Internat. Inst. for Cons.-Can. Group* 3 (Spring 1978), 7-12 (AATA, 16-493).

L. Masschelein-Kleiner, *Liants, Vernis, et Adhésifs Anciens* (Brussels, 1978).

P. Mitanov and V. Todorov, "Compatibility Among Resins, Waxes, and Polymers, Applied to Conservation and Restoration Techniques," *ICOM Comm. for Cons.* (Zagreb, 1978), paper 78/17/3, 1-10. In French (AATA, 16-1854).

1979   R.H. Lafontaine, "Decreasing the Yellowing Rate of Dammar Varnish Using Antioxidants," *Studies in Conservation* 24 (1979), 14-22 (AATA, 16-1492).

1980   G. Hedley, "Solubility Parameters and Varnish Removal: A Survey," *The Conservator*, no. 4 (1980), 12-18 (AATA, 17-1342).

K. Raft, "An Examination of the Value of the Re-forming Technique in Practice," *Studies in Conservation* 25 (1980), 137-140 (AATA, 18-478).

G. Torraca, *Solubilité et Solvants Utilisé pour la Conservation des biens Culturals* (Rome, 1980). In French (In Spanish, 1982; In English, 1984).

1981   E. DeWitte, "The Use of Modern Synthetic Materials for Conservation and Restoration, New Results," *Mitteilungen Deutscher Restaurtoren-Verband* (1981), 46-48. In German (AATA, 20-91).

E. DeWitte, M. Goessens-Landrie, E. J. Goethals, K. Van Lerberghe, and C. Van Springel, "Synthesis of an Acrylic Varnish with High Refractive Index," *ICOM Comm. for Cons.* (Ottawa, 1981), paper 81/16/4, 1-7 (AATA, 18-1382).

R. L. Feller, "Developments in the Testing and Application of Protective Coatings," *ICOM Comm. for Cons.* (Ottawa, 1981), paper 81/16/1, 1-6 (AATA, 18-1386).

R. L. Feller, M. Curran, and C. Bailie, "Photochemical Studies of Methacrylate Coatings for the Conservation of Museum Objects," in *ACS Symp. Ser. 151* (Photodegradation and Photostabilization of Coatings) (1981), 183-196 (AATA, 18-1387).

R. H. Lafontaine, "Use of Stabilizers in Varnish Formulations," *ICOM Comm. for Cons.* (Ottawa, 1981), paper 81/16/5, 1-17 (AATA, 18-1411).

L. Masschelein-Kleiner and P. Taets, "Contribution to the Study of Natural Resins in the Art," *ICOM Comm. for Cons.* (Ottawa, 1981), paper 81/16/3, 1-8 (AATA, 18-1418).

L. Masschelein-Kleiner, *Les Solvants* (Brussels, 1981). In French.

C. Sease, "The Case Against Using Soluble Nylon in Conservation Work," *Studies in Conservation* 26 (1981), 102-110 (AATA, 18-1704).

R. White, "A Review with Illustrations of Methods Applicable to the Analysis of Resin/Oil Varnish Mixtures," *ICOM Comm. for Cons.* (Ottawa, 1981), paper 81/16/2, 1-9 (AATA, 18-1455).

1982 T. Sabita, "Diffusion of Sulfur Dioxide and Chlorine Through Poly(Vinyl Acetate) Films," *Birla Archaeological and Cultural Research Institute Research Bulletin* 4 (1982), 35-41 (AATA, 20-1161).

H. Schnabel, "Detrimental Effect of PVC on Coins," *Neue Museumkunde* 25 (1982), 139-141. In German.

1983 S. M. Blackshaw and S. E. Ward, " Simple Tests for Assessing Materials for Use in Conservation," in *Resins in Conservation*, eds. J. O. Tate, N. H. Tennent, and J. H. Townsend (Edinburgh, 1983), 2-1 to 2-15 (AATA, 19-953).

J. Ciabach, "Investigation of the Crosslinking of Thermoplastic Resins Affected by UV Radiation," in *Resins in Conservation*, eds. J. O. Tate, N. H. Tennent, and J. H. Townsend (Edinburgh, 1983), 5-1 to 5-8 (AATA, 19-965).

E. DeWitte, "Resins in Conservation—Introduction to Their Properties and Applications," in *Resins in Conservation*, eds. J. O. Tate, N. H. Tennent, and J. H. Townsend (Edinburgh, 1983), 1-1 to 1-6 (AATA, 19-979).

C. V. Horie, "Reversibility of Polymer Treatments," in *Resins in Conservation*, eds. J. O. Tate, N. H. Tennent, and J. H. Townsend (Edinburgh, 1983), 3-1 to 3-6 (AATA, 19-996).